Water Contamination Emergencies
Monitoring, Understanding and Acting

Water Contamination Emergencies

Monitoring, Understanding and Acting

Edited by

K. Clive Thompson
Alcontrol Laboratories, Rotherham, UK

Ulrich Borchers
IWW Rheinisch-Westfaelisches Institut fuer Wasser, Muellheim an der Ruhr, Germany

RSC Publishing

The proceedings of the International Conference on Water Contamination Emergencies: Monitoring, Understanding, Acting, held at the IWW Water Centre, Muelheim an der Ruhr, Germany on 11th - 13th October 2010.

Special Publication No. 331

ISBN: 978-1-84973-156-0

A catalogue record for this book is available from the British Library

Published by The Royal Society of Chemistry,
Thomas Graham House, Science Park, Milton Road,
Cambridge CB4 0WF, UK

Registered Charity Number 207890

For further information see our web site at www.rsc.org

Printed and bound in Great Britain by CPI Antony Rowe, Chippenham and Eastbourne

PREFACE

This Water Contamination Emergencies Conference (WCEC) held at Muelheim in Germany was the fourth in the series and was the most ambitious. The main session themes running through the three-day programme were Monitoring, Understanding and Acting. This was followed by a final session on Case Studies and Lessons Learnt. Competently reacting to water contamination emergencies involving low probability/high impact contamination events (chemical, biological or radioactive) in source waters or treated water is essential to ensure a timely and effective response that significantly minimises or avoids adverse impacts on consumers and the environment.

Although the question raised in the first WCEC conference, "Can we cope?" was answered in a positive way, there were a number of "Yes, buts". Some of these "buts" were addressed in the second conference when ways of "Enhancing our response" were considered by identifying areas of good practice and introducing heightened security within a CBRN context. Then the "Collective responsibility" event was held in April 2008 and concluded that, in general, water companies and key agencies respond well and were well positioned to deal with typical emergencies, but more awareness and pre-planning could greatly improve the response.

For this latest event, prevention and response strategies were outlined. Some interesting developments concerning novel on-line monitoring systems were described in detail. For very long river systems such as the Rhine, reliable early warning of any impending contamination which poses a risk to the drinking water production is essential. Other areas covered included how water companies prepare a strategy for unusual or unpredictable emergency incident situations; ensuring full appreciation of the impact and ramifications of emergency contamination incidents; how to efficiently handle these incidents; the importance of efficient communications ensuring that all relevant organisations and stakeholders are kept fully updated; the need for good mutual aid arrangements between water companies and the importance of communicating with consumers throughout incidents and being fully aware of their perceptions. Another often overlooked issue is the need for robust remediation strategies.

During this WCEC4 conference, it became very clear that in preparing for and responding to a water contamination emergency, a considerable amount of effort and detailed planning is required to achieve a desirable outcome. Also regular updating of any plans is essential. One conclusion was that water companies will never stop learning how to improve their response to major incidents. These companies cannot afford to provide an unfit for purpose response to a major incident if they are to maintain the confidence of their customers.

The 19 chapters of this book have been produced by experts in the field and they outline various aspects of the above issues. A fifth conference entitled "Water Contamination Emergencies V: - Managing the Threats" is to be held at IWW in Muelheim from 5th to 7th November 2012. Full details will be available at www.WCEC5.eu in June 2011.

Prof. K. Clive Thompson
Dr Ulrich Borchers
January 2011

Contents

DRINKING WATER SAFETY: GUIDANCE TO HEALTH AND WATER PROFESSIONALS – AND OTHER HEALTH PROTECTION ISSUES ON WATER SAFETY

V. Murray and G. Lau

Centre for Radiation, Chemical and Environmental Hazards, Health Protection Agency, 2nd Floor, 151 Buckingham Palace Road, London SW1W 9SZ, UK

1 INTRODUCTION

The water industry and all its partners are key to the provision of wholesome drinking water to all within our populations. Part of this process is to work with water industry regulators and public health professionals who are concerned for the health of the people. This international conference provides a valuable opportunity to facilitate this and share knowledge by concentrating on water contamination emergencies and examining three key issues of monitoring, understand and acting.

Chemical incidents are not infrequent and may occur as very rapidly obvious releases, such as chemical spills, fires and explosions, or as less immediately apparent events such as contamination of a food product or land contamination. Incidents can occur accidentally or deliberately.

Some chemical incidents may have an impact beyond their original location, in some cases crossing national borders. For example, in north-west Romania cyanide was released from a gold mine into the local river system, leading to fish deaths in three countries.[1] More recently a chemical sludge spill in Hungary on 4 October 2010 has required acute public health impact assessment following the eight deaths and over 120 injured. The immediate health effects of the spill included drowning and chemical burns due to an elevated pH (>12) of the red sludge. Ongoing assessment is needed of the health effects of possible exposure to dust, water and locally produced food that may contain increased amounts of heavy metals. The sludge entered the river Danube and may spread in attenuated forms to countries downstream; this possibility and related health effects will be evaluated. While serious short-term health effects are considered unlikely, potential medium- and long-term effects through contamination from heavy metals (for example, entering the food chain) requires ongoing assessment.[2]

Chemical incidents that lead to human exposure present an important public health challenge both nationally and globally. As part of this, the Drinking Water Safety: Guidance to Health and Water Professionals provides a tool to assist in managing water contamination emergencies in England and Wales. This guidance has been developed by the Drinking Water Inspectorate (DWI) and Health Protection Agency (HPA) and was published in 2009.

This chapter will consider what the roles of the DWI and HPA are in managing water contamination emergencies, the joint guidance and look at chemical incident response and the water shortages arising from flooding in practice. In addition, it will look at research in development, such as the UK Recovery Handbook for Chemical Incidents and address where the work of the water industry and international issues can contribute together towards areas where there are global initiatives such as the Millennium Development Goals.

2 DRINKING WATER INSPECTORATE

The DWI is the drinking water quality regulator for England and Wales. It was formed in 1990 on the privatisation of the water industry. It is part of the Department for Environment, Food and Rural Affairs (Defra), but its Chief Inspector is appointed by the Secretary of State for Environment, Food and Rural Affairs (in England) and National Assembly for Wales, and acts independently of government.

The overarching objective of the DWI is to maintain public confidence in the safety and quality of public water supplies through exercise of its powers of reporting, audit, inspection, enforcement and prosecution. In addition, drinking water inspectors are engineers/scientists with considerable water supply and water quality monitoring experience, therefore the DWI also has a role in providing government with advice on water supply and quality matters. The regulatory framework for water supplies in England and Wales is set out in the Water Industry Act 1991 (the 1991 Act). The 1991 Act was amended by the Water Act 2003. The 1991 Act defines the powers and duties under which the DWI operates and also the duties of water companies and licensees. Under the 1991 Act the authorities responsible for regulating the quality of public supplies are the Secretary of State for Environment, Food and Rural Affairs (in England) and National Assembly for Wales.[3]

3 HEALTH PROTECTION AGENCY

The HPA identifies and responds to health hazards and emergencies caused by infectious disease, hazardous chemicals, poisons or radiation.[4] It gives advice to the public on how to stay healthy and avoid health hazards, provides data and information to government to help inform its decision making and advises people working in healthcare. It also makes sure the nation is ready for future threats to health that could happen naturally, accidentally or deliberately.

The HPA combines public health and scientific knowledge, research and emergency planning within one organisation and works at international, national, regional and local levels. It also supports and advises other organisations that play a part in protecting health. The HPA's advice, information and services are underpinned by evidence-based research. It also uses its research to develop new vaccines and treatments that directly help patients.

Although set up by government, the HPA is currently independent and provides advice and information that is necessary to protect people's health. It exists to help protect the health of everyone in the UK; the ambition is to lead the way by identifying, preparing for and responding to health threats.

In the Health Protection Agency Strategic Overview 2010-2015, Preparing, Preventing, Responding,[5] the HPA's functions are summarised as:

Reducing key infections.

Minimising the health impact of environmental hazards including radiation, chemicals, poisonings and extreme events such as flooding.

Supporting safe and effective biological medicines.

In response to the exponential increase in international travel and trade, and emergence and re-emergence of international disease threats and other health risks, 194 countries across the globe have agreed to implement the International Health Regulations (2005).[6] This binding instrument of international law entered into force on 15 June 2007. The regulations require Member States to strengthen core surveillance and response capacities at primary, intermediate and national levels, as well as at designated international ports, airports and ground crossings. The HPA acts as the focal point for these regulations in the UK.[7]

It is likely that the HPA will become core to a new Public Health Service and join this in 2012.[8]

Every day in the UK, HPA advice is sought in response to chemical incidents. The HPA provides authoritative scientific and medical advice to the NHS and other bodies about the known health effects of chemicals, poisons and other environmental hazards.[9] Of the 967 incidents recorded between 1 January – 31 December 2009,16% (154) of acute chemical incidents resulted in evacuation of the nearby population. The chemical group most frequently identified was products of combustion (31%, 315) with the majority being designated as fires. This is followed by "other organic" chemicals (14%, 133) and "other inorganic" chemicals (10%, 94).[10]

The HPA has developed resources and services for the management of water related chemical incidents including:

Incident checklists,[11] which the HPA currently provides for flooding incidents, odour complaints and other water related chemical incidents. These checklists are designed as an aide-memoire for public health professionals and other emergency responders.

The Chemical Hazards and Poisons Report,[12] which is published for staff in the HPA, NHS, government departments and allied organisations, as well as first line responders such as fire and ambulance services. It is also useful to students and members of the public with an interest in environmental public health.

The Compendium of Chemical Hazards,[13] which is an online information resource for the public and all public health professionals who may be involved in advising and responding to chemical incidents.

4 DRINKING WATER SAFETY: GUIDANCE TO HEALTH AND WATER PROFESSIONALS

This guidance has been developed jointly by the DWI and HPA.[14] It is intended to inform public health and other health professionals about the structure and legal framework of the water industry in England and Wales.

In 2007, there was a major water supply incident involving the loss of water supplies to 160,000 properties in Cheltenham, Gloucester, Tewkesbury and a large part of rural Gloucestershire due to the waterworks being inundated with flood water (see Section 8). Subsequent to this, and other incidents, national level discussions between the DWI and HPA led to an agreement to prepare and publish joint guidance to health and water professionals in support of drinking water quality risk assessments and the issuing of consumer protection advice.

In their day-to-day role, water quality scientists in the water industry work closely with health professionals in the HPA and local authorities. The maintenance of sound working relationships is very important to the delivery of effective and timely responses to water quality incidents and emergencies.

In the preparation of this guidance it was apparent that the safety of drinking water in England and Wales is something the public is able to take for granted, because the day-to-day water supply arrangements in place are comprehensive and demonstrably based on sound science with a fully transparent system of independent scrutiny and appropriate sanctions in place. Accordingly, the guidance contains nothing new and its adoption did not require any special action to be taken by the water industry or health professionals over and above its incorporation into existing training regimes, and its inclusion in water supply and public health operating and emergency management procedures.

The guidance considers the legal framework of the DWI, HPA, public water supplies, private water supplies and that of local authorities. It describes the legal framework of wholesome drinking water which by law (the 1991 Act) drinking water must be wholesome at the time of supply. Wholesomeness is defined by reference to drinking water quality standards and other requirements set out in the Water Supply (Water Quality) Regulations 2000 (as amended) which apply in England and the Water Supply (Water Quality) Regulations 2001 (as amended) which apply in Wales. The guidance notes that many of the standards come from the 1998 European Drinking Water Directive, which came into force fully on 25 December 2003. The Directive focuses on those parameters of importance to human health, but it also includes others that relate to the control of water treatment processes and the aesthetic quality of drinking water. The Directive allows Member States to set additional or tighter national standards to secure the good quality of drinking water already achieved and to prevent it from deteriorating in the future.

Water companies will take a number of actions to protect public health in association with the DWI and HPA. The responsibility for issuing warning notices to consumers and providing alternative water supplies (rezoning, tankers, bowsers and bottles) rests, at all times, with the water company. Precautionary advice is issued by water companies to householders in the form of letters, leaflets or warning notices the public is familiar with, and is therefore responsive to, such advice coming from their water supply company. Depending on a water contamination emergency the water companies may issue three types of warning message:

Boil before use for drinking and food preparation.

Do not use for drinking or cooking.

Do not use for drinking, cooking or washing.

5 LONDON WATER CONTAMINATION INCIDENTS

A local agreement has been made between the HPA and DWI for providing notifications on drinking water related incidents involving health protection issues. The local agreement aims to support the London Early Alerting System.[15] Since July 2007, a total of 44 notifications regarding drinking water related incidents in the London region were received by the HPA from the DWI (Figure 1). The majority of these incidents were located in North East/North Central, followed by North West, South East and South West London areas (Figure 2).

Since more than 40% of London's water mains are over 100 years old,[16] it is not surprising to find that the majority of the incidents were due to burst water mains. Out of the seven incidents that are categorised as chemical contamination, six were related to drinking water contaminated with hydrocarbons with no serious public health consequences for these incidents. The remaining incident concerned elevated levels of ammonia found in a water treatment works, which was resolved before the contamination entered the drinking water supply. Other categories of incidents were mostly temporary i.e. discolouration, heavy rainfall, leakages, taste and odour etc., and with minimal impact on public health.

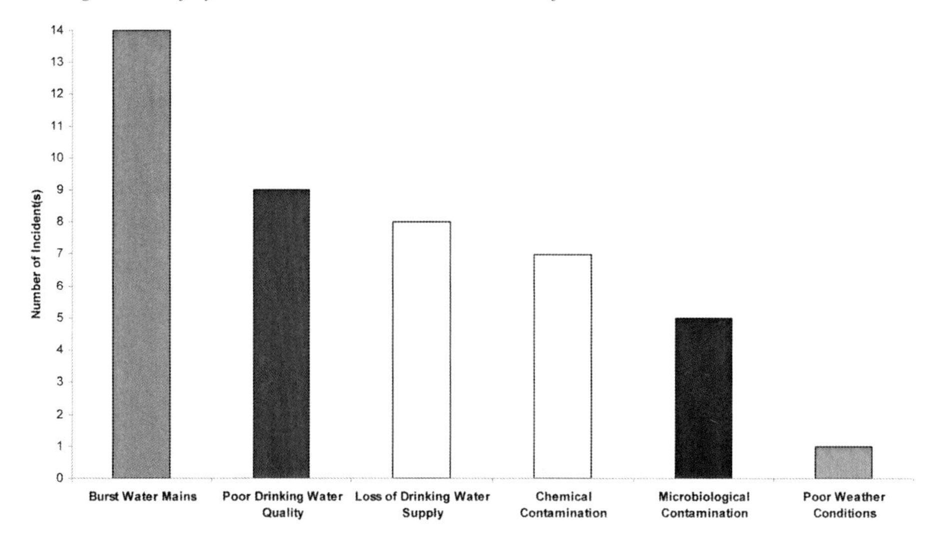

Figure 1 *Categories and occurrences of drinking water related incidents in the London region from July 2007 to June 2010.*

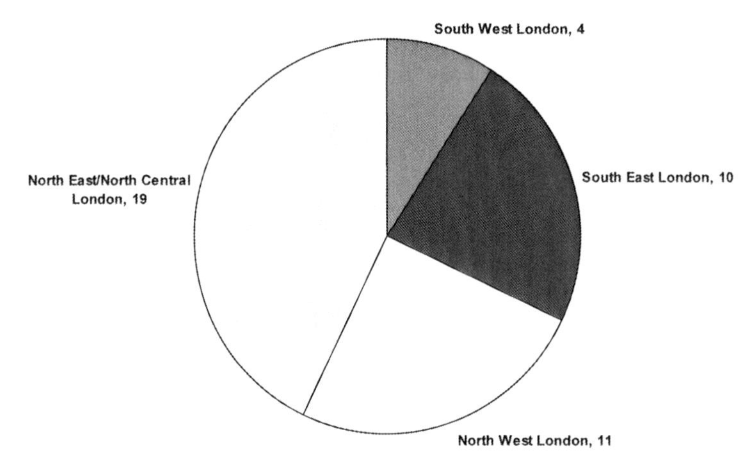

Figure 2 *Notifications for drinking water related incidents in London Region from July 2007 to June 2010.*

6 PUBLIC HEALTH/PROTECTION RESPONSE

If an initial risk assessment indicates a risk to public health, whether actual or potential, then action is required. This will involve immediate care, clinical and health protection support and co-ordination in addition to working with other agencies and organisations in a timely manner. Health protection and public health response professionals act as an interface between clinical health care and overall management of a chemical incident. It is they who will attend the operational, tactical or strategic incident meetings. Most incidents are resolved at the operational level (bronze) requiring public health support by telephone usually, some require tactical support (silver) and a few go to the strategic level (gold).

Occasionally, in the UK, the Science and Technical Cell (STAC)[17] or, perhaps, the Cabinet Office Briefing Room[18] may be required (Figure 3). Public health professionals from the HPA, along with DWI engineers/scientists, may be asked to contribute at STAC and they, with toxicological and other expert knowledge, will contribute to decisions including sheltering and evacuation and the calling in of an incident control team if required.

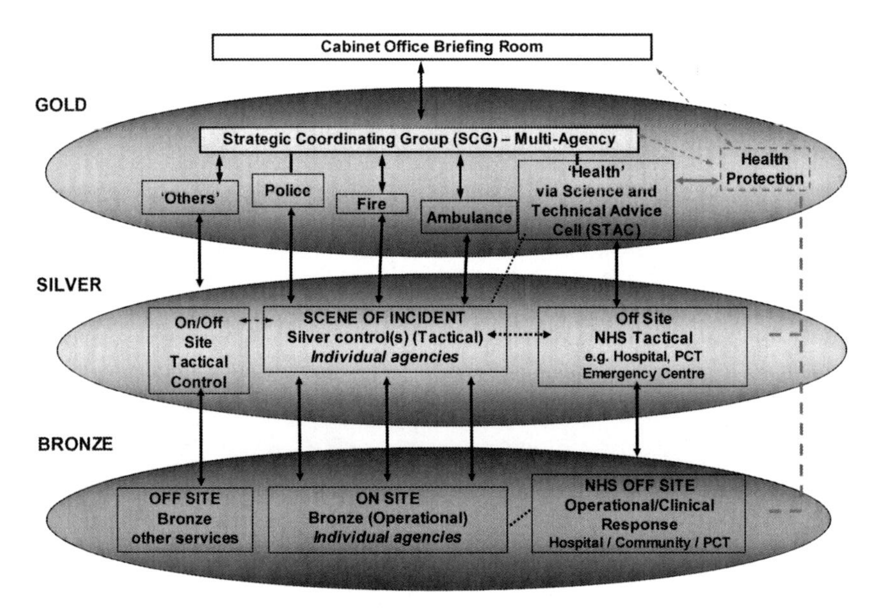

Figure 3 *Command and control levels.*

7 BUNCEFIELD OIL DEPOT FIRE, HERTFORDSHIRE, 2005

In the early hours of 11 December 2005, a number of explosions occurred at Buncefield Oil Storage Depot, Hemel Hempstead, Hertfordshire. Significant damage occurred to both commercial and residential properties in the vicinity and the fire burned for several days. Three Valleys Water responded to the incident and assisted the emergency services by making water available for fire fighting and providing local knowledge to the multi agency team led by the Environment Agency set up to manage the environmental impact of the incident.

Drinking water supplies to consumers in the area were not directly affected by the incident, however, Three Valleys Water took the precautionary step of fitting of non-return valves close to the site (preventing the risk of contamination by backflow of water from the site). The local supply arrangements were also reconfigured with some assets remaining out of supply.

Water used to fight the fire (so called firewater) was collected in bunds on site, and later removed to secure off-site storage. However, some firewater did escape from the site. This water is likely to have contained hydrocarbons and polycyclic aromatic compounds (a group of chemicals commonly found in residues from burning coal, fuel and oil) and residues of fire-fighting foam (which contains zinc and a chemical called perfluorooctane sulphonate or PFOS).[19]

The DWI established an advisory level for PFOS in drinking water of no greater than 3 μg/l at the time. This advice followed consultation with health professionals at the HPA.

The HPA stated that: 'It appears unlikely that a lifetime's consumption of drinking water containing concentrations up to 3 μg/l would harm human health'. In all cases, the detected PFOS levels were below this advisory level.[20]

8 FLOODING AND LOSS OF DRINKING WATER SUPPLIES, GLOUCESTERSHIRE, 2007

The loss of drinking water supplies to parts of Gloucestershire in July and August 2007 following the flooding of the Mythe Water Treatment Works (WTW) was a significant issue. The Mythe WTW is located near to Tewkesbury, Gloucestershire, on the bank of the River Severn close to the confluence with the River Avon. It supplies approximately 160,000 properties in the towns of Cheltenham, Gloucester, Tewkesbury and across a large part of rural Gloucestershire. The treatment process consists of clarification, filtration, ozone, granular activated carbon and chlorination. Treated water is then pumped to Hewletts service reservoir (SR), which supplies Cheltenham; and to Churchdown SR which supplies the remaining areas. These large strategic service reservoirs in turn feed a total of 22 smaller service reservoirs serving the wider parts of the distribution system.

Historically, the treatment works site has been inundated by flood water on a number of occasions, most notably in 1947, but the operation of the works has never previously been lost due to flooding. These historic floods were predominantly caused by a rise in the level of the River Severn. On this occasion the main source of flood water was the River Avon. The 'Flood Emergency Response Plan' for the site detailed the heights at which flood water might breach defences and actions to be taken (predominantly sandbagging buildings and turning the power off). Previous flood events had developed slowly, because of the size of the flood plain but on this occasion the flooding was rapid due to exceptional rainfall over a short period of time. In the two days immediately preceding the event, approximately three months rainfall occurred causing both the River Severn and the River Avon to flood simultaneously. The flood water reached a maximum height greater than the predicted 1 in 150 year's level.

9 HANDBOOKS FOR THE MANAGEMENT OF CHEMICAL AND RADIOLOGICAL WATER INCIDENTS

If a radiological event occurred near an open source of a water supply, then the water would probably pass through an established treatment works prior to being supplied to the consumer. Consequently, any such incident could lead to exposure to radiation for both consumers and the operatives that work in the affected water treatment works. It is important therefore that the drinking water industry has information and guidance to assess the radiological impact on both consumers and operatives. Two complementary handbooks have been produced that provide guidance to the water industry and those decision-makers that would be responsible for managing the response to a radiological incident.

The UK Recovery Handbook for Radiation Incidents[21] contains three Parts which address Food Production Systems, Inhabited Areas, and Drinking Water, respectively. The Drinking Water Part provides guidance on recovery options for reducing doses from the ingestion of drinking water by members of the public. It covers radiological aspects, descriptions of the options and decision trees to assist in the choice of options. In order to help people become familiar with how to use the handbook and to use it effectively, some training aids have been devised; in particular, generic training scenarios with worked examples.

In addition, the Handbook for Assessing the Impact of a Radiological Incident on Levels of Radioactivity in Drinking Water and Risks to Operatives at Water Treatment Works[22] has been produced to provide a tool for the water industry to manage the potential risks to operatives working in a treatment works. It can be used to help the water industry make decisions on how the works can be operated in the event of a radiological incident and to manage any radiation exposures to the operatives. The handbook covers the effectiveness of drinking water treatment processes in removing radionuclides, a methodology to assess radiation exposures to operatives working within drinking water treatment works and guidance on where radionuclides may concentrate within treatment works and the impact this may have. Worked examples to assist users both in planning for a radiological incident and the management of a radiological incident are included.

Mirroring the success of the UK Recovery Handbook for Radiation Incidents, the HPA is developing a UK Recovery Handbook for Chemical Incidents,[23] in collaboration with the following partners:

Defra
Food Standards Agency
Northern Ireland Environment Agency
Home Office
Scottish Government

Development of the UK Handbook for Chemical Incidents will take three years, the results of which will be released in May 2012. On completion, the product is intended to provide a user-friendly online reference handbook in PDF format, which will aid all relevant government departments, agencies, local authorities and other stakeholders involved in the recovery phase of a chemical incident. It will mirror the UK Handbook for Radiation Incidents as far as possible (covering Food Production Systems, Inhabited Areas and Water) but the Water Part will address surface waters, groundwaters and marine waters as well as drinking water.

There are thousands of different chemicals which could be potentially involved in a chemical incident. The UK Recovery Handbook for Chemical Incidents considers all chemicals using a generic approach and identifying the important physicochemical characteristics from the point of view of recovery options.

The main aim of the project is to develop a framework for choosing an effective recovery strategy and a compendium of management options soundly based on science, taking into account acceptable, practicable and achievable practices through the involvement of various stakeholders. It will contain decision trees, checklists and descriptions of the recovery options, following the format of the UK Recovery Handbook for Radiation Incidents.

10 MILLENIUM DEVELOPMENT GOALS AND OTHER INTERNATIONAL INITIATIVES

The United Nations Millennium Declaration of 2000[24] incorporated eight Millennium Development Goals (MDGs) designed to address basic human needs and rights for every individual around the world. All 192 United Nations Member States agreed to the contents of the declaration, which called for the achievement of these goals by year 2015. The eight Millennium Development Goals are:

1. Eradicate extreme poverty and hunger
2. Achieve universal primary education
3. Promote gender equality and empower women

4. Reduce child mortality rate
5. Improve maternal health
6. Combat HIV/AIDS, malaria, and other diseases
7. Ensure environmental sustainability
8. Develop a global partnership for development

The global assessment of water supply and sanitation forms part of MDG 7, which aims to provide:

Access to water, defined as being '*the availability of at least 20 litres per person per day from a source within one kilometre of the users' dwelling*'

An improved drinking water source, defined as '*a drinking water source or delivery point that, by nature of its construction and design, is likely to protect the water source from outside contamination, in particular from faecal matter*'

Overall water access by 2015 is on target, but the sustainable development and efficient use of water resources remains a global target.

The Global Platform for Disaster Risk Reduction: Second Session, was held from 16 to 19 June 2009 in Geneva, Switzerland. It recognised that the world is increasingly facing threats from natural disasters, with the impacts of climate change compounding the situation. In the Chair's summary in the closing plenary, specific targets were also identified, reflecting the conference's deliberations and as catalysts for cutting deaths and economic losses brought on by disasters.[25]

As a result of the Global Platform the World Health Organization (WHO) and the United Nations Secretariat for International Strategy for Disaster Reduction (UNISDR) launched the Thematic Platform: Disaster Risk Reduction for Health at the International Day for Disaster Reduction on 14 October 2009. They have committed to establish the platform, through which local, national and international partners will collaborate on actions to reduce deaths, injuries and illness from emergencies, disasters and other crises (WHO and UNISDR 2009). Of note the WHO and UNISDR state that we all face enormous challenges from natural hazards, conflicts, food crises, climate change, disease outbreaks and now pandemic influenza. The historical record should be sufficient reason to increase investment in disaster risk reduction. However, risks are expected to increase due to factors such as climate change affecting the frequency and severity of hazards and settlement of risk-prone areas due to urbanisation pressures.

In recognition of the links between disaster risk reduction and climate change adaptation, the Intergovernmental Panel on Climate Change (IPCC) with UNISDR has undertaken a special report on Managing the Risks of Extreme Events and Disasters to Advance Climate Change Adaptation. This report, due out in 2011, will help policy-makers evaluate options for reducing disaster risks related to climate change. It will also be included in the IPCC's Fifth Assessment Report.

11 CONCLUSION

The HPA routinely responds to chemical, biological, radiological and extreme events enquiries, incidents and outbreaks through:

Preparing
Preventing
Responding

The HPA therefore identifies health related issues and undertakes research to respond to these issues with local, national and international partners.

The HPA works closely with the DWI and this is exemplified by the joint guidance Drinking Water Safety: Guidance to Health and Water Professionals published in 2009.

With impact of climate change and extreme natural events such as flooding, the need to improve water quality methodology by tools such as those developed between the DWI and HPA are of value.

References

1. United Nations Environment Programme, 2000. *Cyanide spill at Baia Mare Romania: UNEP / OCHA Assessment Mission.* [Online] Geneva: United Nations Environment Programme. Available at: http://www.reliefweb.int/rw/RWFiles2000.nsf/FilesByRWDocUNIDFileName/ACOS-64CHS2-baiamare.pdf/$File/baiamare.pdf [Accessed 3 November 2010].
2. World Health Organisation Regional Office for Europe, 2010. *Hungary: WHO/Europe to assess health impact of sludge spill in Hungary.* [Online] (Updated 12 October 2010) Available at: http://www.euro.who.int/en/where-we-work/member-states/hungary/sections/news/2010/10/whoeurope-to-assess-health-impact-of-sludge-spill-in-hungary [Accessed 3 November 2010].
3. Drinking Water Inspectorate, 2009. *About us.* [Online] (Updated 1 July 2010) Available at: http://www.dwi.gov.uk/about/index.htm [Accessed 3 November 2010].
4. Health Protection Agency, 2010. *About the HPA.* [Online] Available at: http://www.hpa.org.uk/AboutTheHPA/ [Accessed 3 November 2010].
5. Health Protection Agency, 2010. *Health Protection Agency Strategic Overview 2010-2015 Preparing, preventing, responding.* [Online] London: Health Protection Agency. Available at: http://www.hpa.org.uk/web/HPAwebFile/HPAweb_C/1274087971865 [Accessed 3 November 2010].
6. World Health Organization, 2008. *International Health Regulations (IHR).* [Online] Available at: http://www.who.int/ihr/9789241596664/en/index.html [Accessed 3 November 2010].
7. Health Protection Agency, 2009. *National Focal Point function.* [Online] (Updated 15 September 2009) Available at: http://www.hpa.org.uk/Topics/InfectiousDiseases/InfectionsAZ/InternationalHealthRegulations/ihr_NFPfunction/ [Accessed 3 November 2010].
8. Department of Health, 2010. *Liberating the NHS: Reports of the arms-length body review.* [Online] London: Department of Health. Available at: http://www.dh.gov.uk/prod_consum_dh/groups/dh_digitalassets/@dh/@en/@ps/documents/digitalasset/dh_118053.pdf [Accessed 3 November 2010].
9. Health Protection Agency, 2010. *Chemicals & Poisons.* [Online] Available at: http://www.hpa.org.uk/webw/HPAweb&Page&HPAwebContentAreaLanding/Page/1153386734384?p=1153386734384 [Accessed 3 November 2010].
10. Health Protection Agency, 2009. *Centre for Radiation, Chemical and Environmental Hazards: Chemical Surveillance Report 1st January – 31st December 2009.* [Online]

London: Health Protection Agency. Available at: http://www.hpa.org.uk/web/HPAwebFile/HPAweb_C/1284475648621
[Accessed 3 November 2010].

11. Health Protection Agency, 2010. *Incident checklists*. [Online] (Updated 14 April 2010) Available at: http://www.hpa.org.uk/ProductsServices/ChemicalsPoisons/ChemicalRiskAssessment/ChemicalIncidentManagement/IncidentChecklists/
[Accessed 3 November 2010)].
12. Health Protection Agency, 2010. *Chemical hazards and poisons report*. [Online] (Updated August 2010) Available at: http://www.hpa.org.uk/Publications/ChemicalsPoisons/ChemicalHazardsAndPoisonsReports/
[Accessed 3 November 2010].
13. Health Protection Agency, 2010. *Compendium of chemical hazards*. [Online] Available at: http://www.hpa.org.uk/Topics/ChemicalsAndPoisons/CompendiumOfChemicalHazards/
[Accessed 3 November 2010].
14. Drinking Water Inspectorate, 2009. *Drinking Water Safety: Guidance to health and water professionals*. [Online] London: Drinking Water Inspectorate. Available from: http://www.hpa.org.uk/web/HPAwebFile/HPAweb_C/1252660062619
[Accessed 3 November 2010].
15. Cordery, R. Mohan, R. & Ruggles, R., 2007. Evaluation of the London Chemical Incident Early Alerting System: (1) an audit of chemical incident reporting in London, two years on, In: V. Murray, ed. 2007. *Chemical Hazards and Poisons Report*. [Online] May, Issue 9. Available at: http://www.hpa.org.uk/web/HPAwebFile/HPAweb C/1194947352137
[Accessed 3 November 2010].
16. Thames Water Utilities Limited, 2001. *Why we are replacing pipes*. [Online] Available at: http://www.thameswater.co.uk/cps/rde/xchg/corp/hs.xsl/2690.htm
[Accessed 3 November 2010].
17. Cabinet Office, 2007. *Provision of scientific and technical advice in the strategic co-ordination centre: Guidance to local responders*. [Online] London: Cabinet Office. Available at: http://www.cabinetoffice.gov.uk/media/132949/stac_guidance.pdf
[Accessed 3 November 2010].
18. Cabinet Office, 2010. *Responding to emergencies: The UK central government response – concept of operations*. [Online] London: Cabinet Office. Available at: http://www.cabinetoffice.gov.uk/media/349120/conops-2010.pdf
[Accessed 3 November 2010].
19. Drinking Water Inspectorate, 2006. *Drinking water 2005: Part 3 – Thames region*. [Online] London: Drinking Water Inspectorate. Available at: http://www.dwi.gov.uk/about/annual-report/2005/Part%203%20-%20Thames%20region.pdf
[Accessed 3 November 2010].
20. Buncefield Investigation, 2008. *The Buncefield Incident 11 December 2005: The final report of the Major Incident Investigation Board – Volume 2a*. [Online] Available at: http://www.buncefieldinvestigation.gov.uk/reports/volume2a.pdf
[Accessed 3 November].
21. Health Protection Agency, 2009. *UK Recovery Handbooks for Radiation Incidents: 2009*. [Online] (Updated 19 April 2010) Available at: http://www.hpa.org.uk/web/HPAweb&HPAwebStandard/HPAweb_C/1259152442006
[Accessed 3 November 2010].

22. Health Protection Agency, 2008. *Handbook for Assessing the Impact of a Radiological Incident on Levels of Radioactivity in Drinking Water and Risks to Operatives at Water Treatment Works*. [Online] (Updated 1 September 2008) Available at: http://www.hpa.org.uk/Publications/Radiation/HPARPDSeriesReports/HPARPD040/ [Accessed 3 November 2010].
23. Galea, A. Brooke, N. Baker, D. Dobney, A. Mobbs, S. & Murray, V., 2010. A UK recovery handbook for chemical incidents, In: V. Murray, ed. 2010. *Chemical Hazards and Poisons Report*. [Online] January, Issue 16. Available at: http://www.hpa.org.uk/web/HPAwebFile/HPAweb_C/1263812796194 [Accessed 3 November 2010].
24. United Nations, 2000. *United Nations Millennium Declaration*. [Online] New York: United Nations. Available at: http://www.un.org/millennium/declaration/ares552e.pdf [Accessed 3 November 2010].
25. Global Platform for Disaster Risk Reduction, 2009. *Outcome document: Chair's summary of the second session*. [Online] Geneva: Global Platform for Disaster Risk Reduction. Available at: http://www.preventionweb.net/files/10750_GP09Chairs Summary.pdf [Accessed 3 November 2010].

A NOVEL APPROACH FOR EARLY WARNING OF DRINKING WATER CONTAMINATION EVENTS

B.H. Tangena[1], P.J.C.M. Janssen[1], G. Tiesjema[1], E.J. van den Brandhof[1], M. Klein Koerkamp[2], J.W. Verhoef[2], A. Filippi[3], W. van Delft[4]

[1] National Institute for Public Health and the Environment (RIVM), Bilthoven, The Netherlands
[2] Optiqua, Enschede, The Netherlands, Singapore
[3] Philips Research, Eindhoven, The Netherlands
[4] Vitens, Leeuwarden, The Netherlands

1 INTRODUCTION

Quality and safety of drinking water is intensively guarded by water companies and authorities, but a 100% safeguard against accidental or intentional contamination is virtually impossible. In case of a public health threatening contamination event in the distribution network, early warning is essential. The relatively fast spreading of contaminants throughout the distribution system implies that the processing time of traditional sampling is often too long. Water companies are looking for ways to continuously monitor, either in-line or on-line, the quality and safety of water in their distribution networks. Such an early warning system is supplementary to the spot sampling and continuous monitoring of the treatment process. A sensor as part of an early warning system should be generic, it should alert on harmful contaminants and it should distinguish between accidents and normal quality fluctuations.

The research presented in this paper has been executed in the project 'Aqua Vitaal: online sensor for the detection of drinking water incidents'. Aqua Vitaal is a consortium of knowledge institute RIVM (The Netherlands), research company Philips Research (The Netherlands), drinking water company Vitens (The Netherlands) and sensor developer Optiqua (Singapore and The Netherlands).

The assessment and validation of sensors for deployment in an early warning system is driven by the ability to detect concentration levels of contaminants down to threshold levels that do not pose short term public health effects. For many substances, standards for long term exposure (lifetime) are available, for instance the drinking-water guidelines of the World Health Organization (WHO) and the drinking-water standards of the European Union. However, specific guideline threshold values for acute exposure in case of an incident are lacking. In the 3rd edition of the WHO Guidelines for Drinking-water Quality[1] this need for emergency values is explicitly noted. On the basis of toxicological literature, short term exposure limits have been derived, denoted as Drinking water Alert Levels (DALs). In accordance with the method as used for deriving US Environmental Protection Agency (EPA) Acute Exposure Guideline Levels for air (AEGLs), the DALs have been derived for three health effect categories: 'no adverse effects', 'serious effects' and 'lethal effects'.

The DALs are used in the evaluation of a new sensor platform based on a Mach Zehnder Interferometer (MZI) concept as developed by Optiqua (chapter 3). This sensor has been tested in laboratory batch experiments. The response to target contaminants for which DALs are derived is analyzed (chapter 4).

Data interpretation software has been developed and applied to distinguish contamination events from 'normal' quality changing patterns (chapter 5). The data algorithm and its application to measured data will be discussed.

2 DERIVATION OF DRINKING WATER ALERT LEVELS

2.1 Background and Definition

'Alert levels' as required for the Aqua Vitaal project, represent the range of concentrations in drinking-water that are relevant from a health-based point of view. To optimally meet this requirement, the templates as used for the Dutch Intervention Values (DIVs) and the US-EPA Acute Exposure Guideline Levels (AEGLs) for inhalation exposure are used[2]. These templates involve deriving estimates of the thresholds above which toxic effects of pre-defined severity can be expected upon single exposure. Thus the whole range of possible toxic effects for any chemical due to its presence in drinking-water is covered, from the first minor signs of exposure up to lethality. As with the DIVs and AEGLs, the Drinking water Alert Levels (DALs) indicate at which levels acute toxic effects in three pre-defined severity categories are to be expected. Similarly like the DIV and AEGL values, the DALs represent an estimate, as precise as possible, of the *threshold* for acute toxic effects of the specified category in humans. This notion of DALs being predicted thresholds applies in particular to the two highest pre-defined effect severity categories, DAL-2 and DAL-3. As with DIVs and AEGLs, for these two values assessment factors used in their derivation should not be higher than strictly necessary. Like the DIVs and AEGLs, these values are intended to be *predictive* rather than *preventive*. The lowest category (DAL-1) is bases on the concept of the Acute Reference Dose (ARfD). The ARfD was chosen here because it is accepted and well known[3].

Deriving DALs requires a review of all available information on acute toxicity of a chemical via the drinking water route. Ideally, this yields an estimate of the threshold for humans, including sensitive subpopulations, for toxic effects in each of the three pre-defined severity categories. As is usual within toxicology, this threshold is expressed as the dose in mg/kg body weight. To convert this body dose to DALs as concentrations in drinking-water, scenarios are used covering single-day exposure (24 hours) via ingestion and showering (inhalation) for an adult and ingestion and bathing (dermal uptake) for a bottle-fed infant.

At present the methodology of DAL derivation as presented in this paper is not widely accepted. One programme where a similar approach already has been applied for drinking water, is the development of 'Provisional Advisory Levels' (PALs) for priority toxic industrial chemicals, chemical warfare agents and pesticides in air and drinking-water, as carried out within US-EPA's National Homeland Security Research Center (NHSRC)[4].

The definition of DALs is as follows:

DAL-1: The DAL-1 is the estimated concentration in drinking-water that can be ingested over a period of 24 h or less without any appreciable health risk to the consumer on the

basis of all known facts at the time of the evaluation. The DAL-1 is based on the concept of the ARfD (Acute Reference Dose), which has been developed for pesticides*.
As is the case for the ARfD, the DAL-1 is based on the most sensitive acute toxic effect for the chemical in question. All adverse effects are taken into account. Thus the critical effect (the adverse effect seen at the lowest dose) may occur in any organ: e.g. in the blood, immune system, nervous system, liver, kidneys or endocrine system. The DAL-1 is a preventive value, incorporating a larger safety margin than DAL-2 and DAL-3.

DAL-2: The DAL-2 is the estimated concentration in drinking-water that can be ingested over a period of 24 h *above* which serious, irreversible or other serious health effects could result among the general population (including all ages and sensitive subpopulations). Examples of severe or irreversible effects possibly occurring above DAL-2 include ocular damage, gastrointestinal irritation/bleeding, organ injury, clinically relevant haemolysis, ocular or dermal effects and pulmonary damage. If toxicity data are insufficient for a chemical, a DAL-2 can sometimes be estimated by a reduction of the respective DAL-3 (depending on the steepness of the dose-response curve).
As follows from the definitions, for concentrations above DAL-1 but below DAL-2 minor and reversible health effects are expected (head ache, minor intestinal complaints).

DAL-3: The DAL-3 is the estimated concentration in drinking-water that can be ingested over a period of 24 h above which life-threatening health effects or lethality in the general population, including all ages and sensitive subpopulations, could occur.

2.2 Toxicological Evaluation

For many chemicals only limited acute oral toxicity data are available, which impacts the derivability of DALs. Ideally, acute toxicity studies with multiple dose levels should be available from which No Observed Adverse Effect Levels (NOAELs) can be derived for DAL-1, -2 and -3 effects, respectively. However, such acute studies are available for relatively few chemicals. Such studies are only routinely carried out for pesticides that are acutely toxic. From these studies, relevant no-effect-levels for DAL-1 and DAL-2 effects can be selected. For many other chemicals however acute oral toxicity has been examined only for the endpoint of lethality. Mostly, this involves determination of the LD_{50}, the dose at which 50% of the animals die. Traditionally, this has been the acute toxicity study carried out routinely in chemical safety testing. In recent years more attention has been given to dose response for the various acute toxic effects a chemical may produce.

When limitations exist in the database of information on acute toxicity, useful surrogate information may be derived from oral studies of longer duration (sub-acute, semi-chronic) carried out with the chemical in question. These studies sometimes provide relevant information concerning acute thresholds of toxicity or they can be used as a cap to delimit the range at which certain acute effects are unlikely to occur. Additionally, acute toxicity via other exposure routes than orally may have been studied more extensively. The results of these studies may be extrapolated to the oral route (see also Solecki et al.[3]). Important toxicological effects used for the DAL-2 and DAL-3 derivation are:

---

* The ARfD is intended as a preventive limit value to be used as a screening tool for possible health risk due to inadvertently high pesticide residues on individual food crops and as such, incorporates a relatively wide safety margin.

- Cholinesterase inhibition
- Carcinogenicity
- Reproductive toxicity
- Lethality

The outcome of the evaluation of acute oral toxicity for animals or humans is the dose that can be used as Point Of Departure (POD) for the three pre-defined effect categories. This POD should quantify the threshold for the selected critical effect. The POD selected refers to the specific sample of healthy laboratory animals or the healthy human volunteers studied in the underlying experiment. For estimating the desired threshold for the entire human population, including potentially sensitive subpopulations, the POD has to be extrapolated to the latter population; this is done by dividing it by one or more assessment factors. This procedure is in agreement with the method used for DIVs and AEGLs. Separate assessment factors are applied for extrapolation from animals to humans (interspecies) and for extrapolation to sensitive (but not hyper susceptible) humans (intraspecies). Since the aim is to estimate *thresholds* for effects, assessment factors should not be higher than strictly needed. In agreement with guidelines for the derivation of DIVs, two extrapolation factors, each ranging from 1 to 10, are used for inter- and intraspecies extrapolation, respectively.

As a default, a factor of 3 will be used. Higher or lower factors are used (case-by-case) depending on the available data. Lower factors than the default may be warranted based on considerations concerning inter/intraspecies differences in the mechanism of action for the chemical in question. Additional assessment factors may be included for extrapolation to a NOAEL when a NOAEL has not been identified (up to 10, depending on the steepness of the dose response curve), for the use of sub-acute/sub-chronic data (<1, with reservations, since it is not known how much lower than 1 may be justified) or for uncertainty in the database (up to 10). If factors move estimated thresholds to implausibly low levels when judged against other evidence from the available database, this indicates that lower factors are warranted. On the whole, the DAL-1, -2 and -3 derived for any compound should credibly reflect available dose response information for that compound, including a reasonable spacing between the estimated thresholds for the three effect categories.

2.3 Exposure Scenarios

For the 24 hours exposure to drinking water three uptake routes are distinguished: drinking, showering (inhalation) and bathing (dermal exposure). These scenarios are summarized in Table 1. Because young infants, particularly bottle-fed infants, consume much more drinking-water on a body weight basis than adults, we use a different exposure scenario for this group. Thus, we calculate DALs from the relevant toxicity thresholds expressed as mg/kg body weight for:

- a 70 kg adult drinking 2 litres of water and showering for 10 minutes;
- a 4.5-month old infant of 6.75 kg drinking 1 litre of water and bathing once per exposure day.

Showering will only be included in the formula when it contributes ≥ 20% to the exposure of adults. For bathing the same figure is applied in the calculation of the exposure of infants.

Table 1 *Exposure scenarios*

	Drinking (oral exposure)	Showering (inhalation exposure)	Bathing (dermal exposure)
Adult (70 kg)	2 litres per day	10 minutes with ventilation volume of 1.5 m^3/hour	-
Infant (6.75 kg)	1 litre per day	-	0.346 litres making contact with the skin*

2.4 Results

In Figure 1 the DAL-values for four substances are presented. They represent three pesticides (aldicarb, azinphosmethyl, fenamiphos) and one toxic chemical (sodium cyanide). The values are given on a logarithmic scale for easy comparison between dose levels of different substances. As expected the DAL-values for adults are higher than for infants. In most cases DAL-1 values are more than a magnitude lower than DAL-2 or DAL-3 values which reflects the preventive nature of the underlying limit value for DAL-1, the ARfD. The DAL-values are typically in the order of magnitude of milligrams per liter. Long-term exposure values typically correspond to concentrations in the order of magnitude of micrograms per litre. Hence, acute health hazards typically correspond to concentrations approximately a factor 1,000 higher than long-term exposure limits. For sodium cyanide the toxicological information was insufficient to derive DAL-3 values.

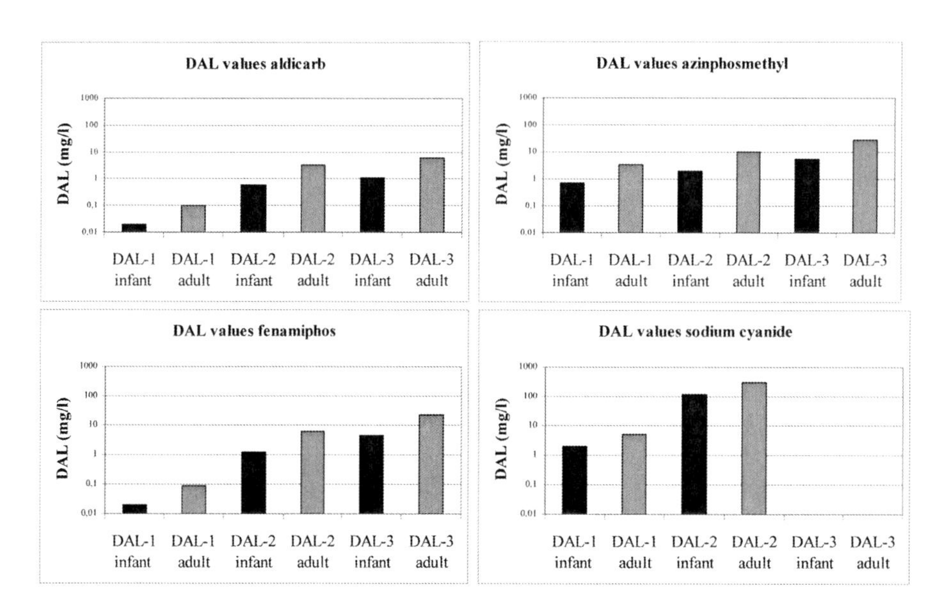

Figure 1 *DAL- values for four target substances*

* This figure is bases on a 0.1 cm layer of water around the exposed skin of a 4.5 month old infant with a body surface of 3460 cm^2.

3 THE OPTIQUA SENSOR

3.1 Characteristics of the Sensor

Optiqua has a generic optical sensor concept that measures minute refractive index changes in water flowing over the surface of the sensor. The Optiqua optical sensor meets four key requirements imposed by an early warning system: rapid detection, generic detection, high sensitivity and low-cost.

Firstly, the sensor's fully continuous measurements allow for instant event detection. In this case an event is defined as an abnormal change in water quality that manifests itself as a change in the water refractive index in a predefined time window.

Secondly, the sensor is responsive to an extensive spectrum of possible contaminants and other substances. The Refractive Index (RI) is a useful generic indicator of water quality. RI is a physical quantity of every substance, defined by the speed of light through that substance, relative to the theoretical speed of light in a vacuum. Any substance, when dissolved in water, will change the refractive index of the water matrix. The change in refractive index is proportional to the concentration and the refractive index of the substance. Virtually all relevant substances have refractive indices that are clearly higher than water. As an example, Figure 2 depicts the concentration of lithium chloride in pure water and the resulting change in refractive index.

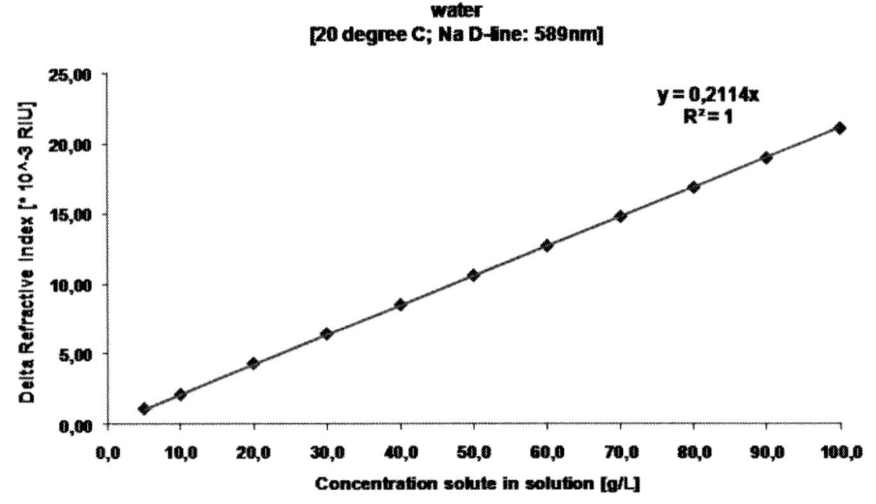

Figure 2 *Index of refractive increment of pure water as a function of the concentration of LiCl (in g per litre). Data taken from the Handbook of Chemistry and Physics*[5].

As can be seen, the relationship between refractive index and concentration is linear. This linearity is maintained when a substance is dissolved in a water mixture containing various elements provided that there is no chemical interaction between the added substance and the elements already present in the initial water solution.

Thirdly, the sensor is highly sensitive. The refractive index as a generic quality indicator for drinking water has been proposed before[6]. However, the sensitivity levels needed for application in drinking water control have not been achieved with commercially available technologies. The Optiqua sensor accurately measures a minimal change in refractive index in the order of magnitude of 10^{-7} RI. Changes of RI in that order of magnitude typically correspond to detection limits of concentration levels in or under the single-digit milligrams-per-liter concentration range. In the example of lithium chloride, the Optiqua sensor has a detection limit that is a factor 1000 smaller than the refractive index of the lowest concentration reported in the Handbook of Chemistry and Physics. Therefore it is important to validate via spiking experiments that the linear relationship between concentration and refractive index still holds for the tested compounds in these low concentration experimental windows.

The practical sensitivity to detect events is not only driven by the detection limit of the sensor, but also by the natural variation present in drinking water. Therefore, in chapter 5 the response is analyzed in relation to a baseline of continuous measurement in regular drinking water flowing from the tap.

Finally, the sensor offers a low cost platform. The sensor is developed in a dipstick probe format (no moving parts) and does not require any reagents. The sensor is suitable for operation in a network at locations with limited or no direct supervision.

3.2 Sensor Design

Optiqua's sensor is based on an integrated optical version of the Mach-Zehnder Interferometer (MZI). The MZI works as an optical scale, measuring differences in refractive index as seen by the sensing arm versus the reference arm.

The basic layout of the MZI consists of an input channel wave-guide that splits up into two identical branches (see Figure 3). After a well-defined length, these two branches are combined again to form the output wave-guide. Light that enters the input wave-guide splits equally over the two branches and combines again at the output wave-guide. The wave-guides are so called buried wave-guides in which the light travelling through the wave-guide is shielded from the environment via a top cladding. By using etching techniques, the top cladding is locally removed at a well-defined position above the wave-guide in the sensing branch. In this so-called sensing window, the evanescent field of the light that travels through the under-laying channel wave-guide, extents into the environment above the sensor and becomes susceptible to changes in refractive index of the water sample on top of the sensor. The resulting change of the effective refractive index leads to a change of the speed of the light in the sensing branch and a change in the relative phase between light that has travelled through the sensing branch as compared to the reference branch. This change in relative phase leads to a change in the interference between light coming from the sensing and reference branch at the combining section and manifests itself as change in the output intensity of the MZI.

The patented Optiqua sensor is an adaptation of the basic MZI design to improve the overall performance in terms of sensitivity, robustness and temperature dependence. By incorporating a modulator section and utilizing a serrodyne modulation concept (see Heideman[7]) an unambiguous phase determination can be performed using Fourier analysis. In addition this concept is robust against laser intensity variations. The integrated optical design with no moving parts further enhances the overall sensitivity and robustness.

The configuration with one sensing window results in a reading that is sensitive for temperature changes. This is explained by both the temperature sensitivity of the bulk refractive index of water and the materials of which the sensor is constructed. In the Optiqua MZI design, the temperature sensitivity is well predictable. Combing the sensor with an independent temperature reading allows for effective compensation for temperature effects. For a more elaborate description of the technological principle, please refer to Heideman[7] and Lambeck[8].

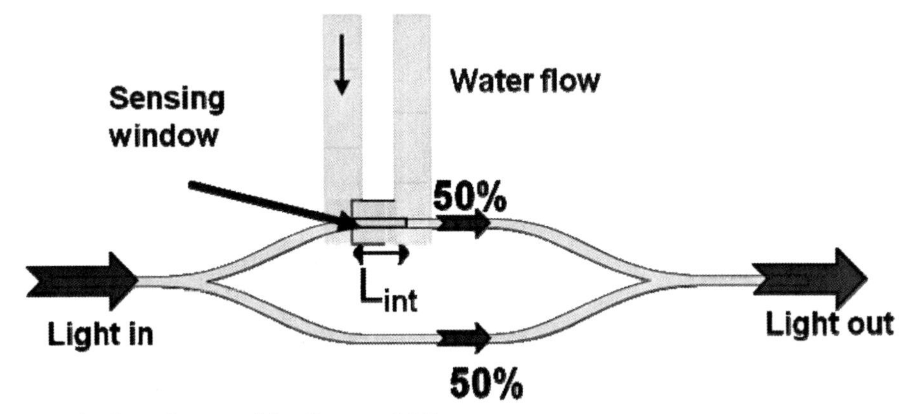

Figure 3 *Basic layout of the Optiqua MZI sensor.*

Changes in refractive index of the water sample on top of the sensor are measured by phase changes of the light propagating over the interaction window. The primary output signal of the sensor is the phase shift of light $\Delta\Phi_{m(easured)}$:

$$\Delta\Phi_m = (2\pi/\lambda)\ L_{int}\ (\partial n_{eff}/\partial n_{water})\ \Delta n_{water} \quad \text{[radians]} \tag{1}$$

where λ is the wavelength of the light in vacuum, L_{int} is the interaction length of the sensing window, $\partial n_{eff}/\partial n_{water} \cong 0.21$ and indicates how sensitive the light travelling through the sensing window is for refractive index changes of the water, and Δn_{water} is the refractive index change in the water flowing over the sensing window.

With λ=850nm and L_{int}=10mm and using equation (1) the change in refractive index of the water Δn_{water} is given by:

$$\Delta n_{water} \cong 4*10^{-4}\ (\Delta\Phi_m/2\pi) \tag{2}$$

4 SPIKING EXPERIMENTS

4.1 Test Setup

The test setup was built around the Optiqua MZI sensor in combination with a flow-through cell (Figure 4). To prevent a surplus of waste, a relatively small volume (1 ml) of test medium was used to test for a change of refractive index. Milli-Q purified water was pumped over the sensor using a micro dialysis syringe pump with a flow rate of 100 µl/minute. The volume of the sample loop was 100 µl. It was loaded using a syringe containing a total volume of 1 ml of the spiked solution. Loading the sample loop with a tenfold higher volume guarantees that the final concentration in the sample loop corresponds to the actual concentration of the prepared spiked test solution. The internal volume of the flow container was 40 µl.

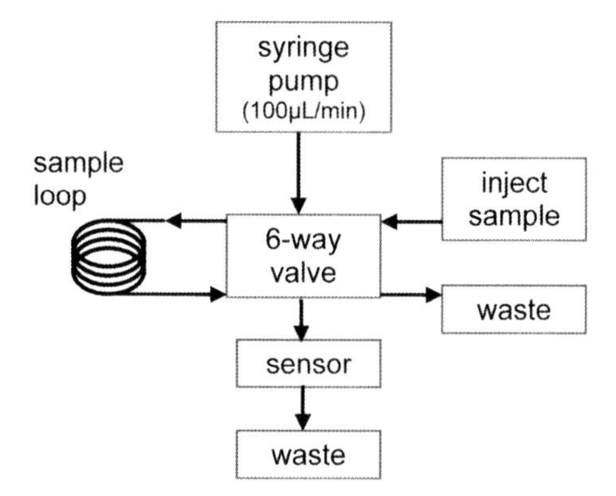

Figure 4 *Test setup*

The total duration of one test run was 6 minutes. During the first three minutes Milli-Q water was pumped over the sensor, thereafter the 6-way sample injector was automatically switched to the sample loop that was pre-loaded with the spiked test solution. After the plug of spiked test solution had passed the flow-cell (approx. 1 minute), purified Milli-Q water flushed again through the flow-cell for another 2 minutes. At the end of the test cycle the 6-way valve was switched to the LOAD position for running the next test in the series.

4.2 Test Compounds and Preparation of Test Solutions

Because the tested compounds are highly toxic, the measurements were performed in a BSL2 cabinet for safety reasons. Room temperature was maintained at 22°C. Chemical characteristics of the test compounds are given in Table 2; the compounds are the same as for the derivation of DAL values.

Table 2 *Tested chemicals in Optiqua flow-through cell*

Chemical	CAS number	Water solubility at 25°C in mg/l	Log K_{ow}	Concentration[1] to be tested (mg/l)
Aldicarb	116-06-3	6030	1.13	0-1.56-3.125-6.25-12.5-25
Azinphos-methyl	86-50-0	2604	0.78	0-1.56-3.125-6.25-12.5-25
Fenamiphos	22224-92-6	20.47	3.23	0-0.625-1.25-2.5-5-10-20
Sodium cyanide	143-33-9	1000000	-1.69	0-1.95-3.91-7.81-15.63-31.25-62.5-125-250-500-1000

1 prepared in Milli-Q purified water

Stock solutions were prepared in Milli-Q water. Test concentrations were prepared at least 24 hours before actual measurement to acclimatize to room temperature. Measurements were done from low to high concentrations. The prepared dilutions were gently homogenized shortly before measurement. The samples to fill the sample loop were drawn about 1 cm under the fluid surface. If the test solutions were measured at different days, they were stored in the dark and at room temperature in between.

The spike test solutions were prepared as follows:
Aldicarb: 6.1 mg was weighted and dissolved in 100 ml Milli-Q water. From this stock the dilutions were prepared with Milli-Q water.
Azinfos-methyl: 5.1 mg was weighted and after 5 minutes of ultra-sonification the chemical compound was completely dissolved in 100 ml Milli-Q water. From this stock the dilutions were prepared with Milli-Q water.
Fenamiphos: 4 mg was weighted and after 2 minutes of ultra-sonification the chemical compound was completely dissolved in 200 ml Milli-Q water. The stock solution had a light yellow color. The color was less profound after 24 hours. This may indicate break down of the chemical.
Sodium cyanide: 40 mg was weighted en dissolved in 20 ml Milli-Q water. From this stock the dilutions were prepared with Milli-Q water.

4.3 Data Evaluation and Results

In Figure 5 the sensor output is given for the complete 6 minutes test runs with different concentrations of Azinphos-methyl. The first 3 minutes of continuous flowing Milli-Q water provide a stable sensor response. The insertion of the spiked sample at levels at 180 seconds leads to a sensor response that is clearly visible in the graph. Higher concentrations of the spiked solution result into a higher response of the sensor.

The step change was quantified by taking the value of the plateau of the sensor response caused by the spiked test sample and subtract the value of the baseline before the spike injection.

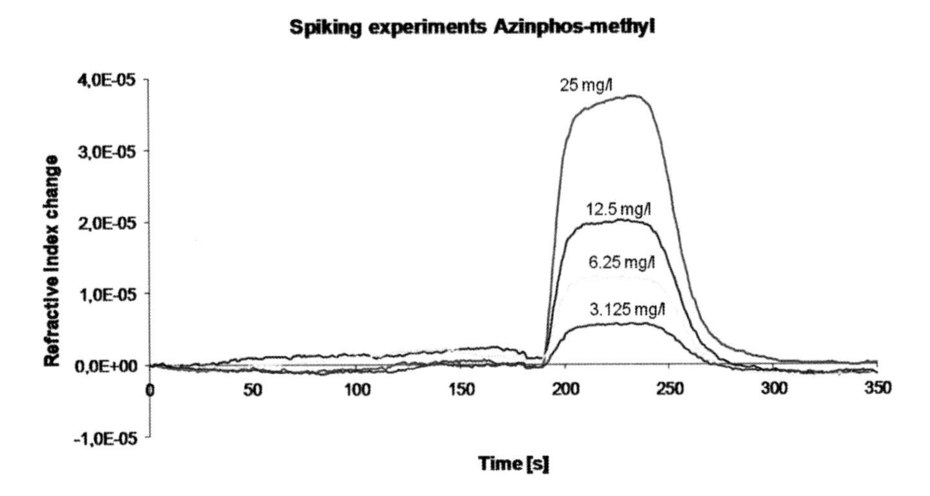

Figure 5 *Sensor response during azinphos-methyl spiking experiment*

Figure 6 depicts the different concentrations included in the experiment and the related step change of the events in Refractive Index (RI) units.

A regression was run to determine the calibration factor:

$$\Delta RI = \beta C \quad (3)$$

with C the concentration of the spiked test solution in mg/l. The linear regression model provides a strong fit with the experimental results (R2 of 99%). The slope β of the regression line can be interpreted as the calibration factor of the sensor response for each mg of the target solution in 1 liter of reference water. For azinphos-methyl the obtained calibration factor is 1.54 * 10-6 (mg/l)-1. We followed the same methodology for the other tested compounds. Table 3 provides an overview of the experimentally derived calibration factors for each of the tested compounds.

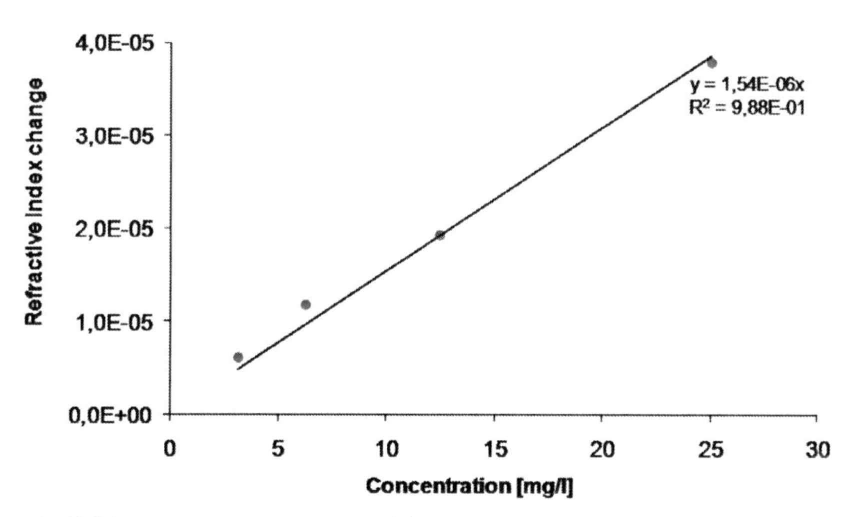

Figure 6 *Calibration curve azinphos-methyl spiking experiment*

Table 3 *Calibration factors and limit of detections*

Chemical	Calibration factor [ΔRI /(mg/l)]	Limit of detection [mg/L]
Aldicarb	$2.0\ 10^{-7}$	1.0
Azinfos-methyl	$1.5\ 10^{-6}$	0.2
Fenamifos	$6.5\ 10^{-7}$	0.3
Sodium cyanide	$1.2\ 10^{-7}$	1.7

Figure 7 depicts the detection limit for the different compounds and compares to the DAL-values. The detection limit was clearly below the DAL-2 values for all these compounds. For azinphos-methyl and sodium cyanide the detection limit was below the DAL-1 value.

The results indicate that the Optiqua sensor is well suited to prevent major health damage from a contamination event by detecting concentrations below the levels of serious and irreversible health damage (DAL-2).

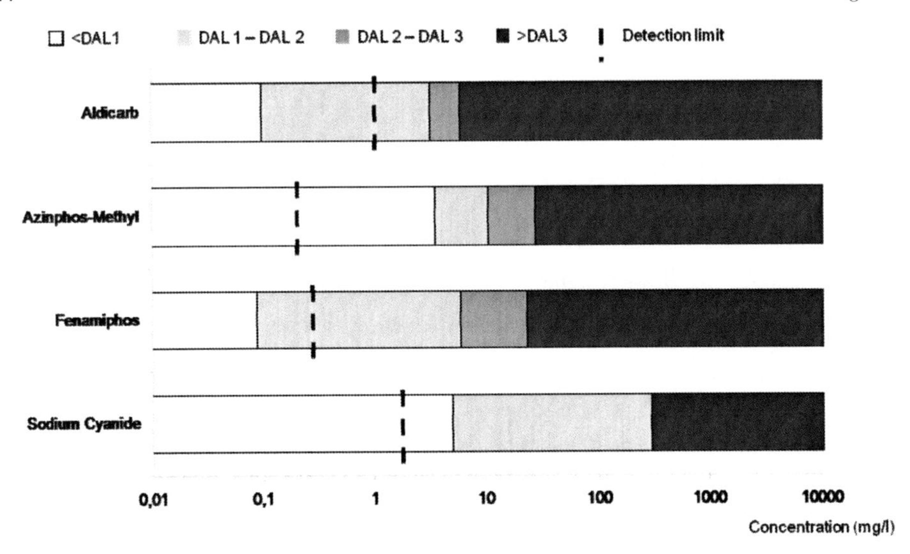

Figure 7 *Detection limit and DAL-values of tested compounds*

5 DATA ANALYSIS

5.1 Background

The data collected by the generic sensor needs to be interpreted to distinguish contamination events from 'normal' water quality changing patterns. The 'normal' quality changes were collected by monitoring the tap water in an office building in The Netherlands and the events by means of the spiking experiments in the lab. The direct observation of the data indicated that the natural variations tend to be smoother than the contamination which generates a sudden change in the sensor response. Although this might not be true if the contamination happens very far from the sensor location, we began developing the event detection algorithms by exploiting this key difference.

From the algorithm perspective, the sensor data are simply a series of digital data. The event detection problem can then be expressed as the detection of a sudden change in a digital data series. This is a well investigated topic in the signal processing community since it can have a multiplicity of applications ranging from modeling of biological phenomena to enhancing picture quality in TV. Quite recently, event detection algorithms have been used to detect contamination events in drinking water distribution systems[9]. The authors present an extensive investigation and provide a rich reference. They apply event detection algorithms to standard water quality indicators such as pH, free chlorine, total organic carbon and specific conductivity. They design event detection algorithms to interpret these sensor readings and provide a way of detecting the contamination event. In this work, we do not apply the detection algorithms to the data of a conventional sensor, but on the data obtained from the new sensor presented in chapter 3 which reacts to a broad spectrum of contaminants. At the same time, this also means that the sensor will be more sensitive to natural variations of water characteristics, thus requiring a proper signal processing stage to distinguish between the contamination events and the natural variations.

The authors propose to use almost conventional edge detection algorithms[10], after having conditioned the sensor signal to improve the quality of the sensor data. The signal conditioning includes a temperature compensation algorithm.

5.2 Temperature Compensation

Since the first sensor data became available, it has been evident that there was a close relationship between the temperature variations of the water and the sensor response.

As we have access to the temperature measurement we can make use of this information and properly compensate for the changes due to temperature variations. By compensating the effects of the temperature, we also limit the detection of events that generates a temperature change. In case of a contamination, the volume of the contaminant is going to be significantly smaller than the volume of the water in the pipes. Therefore, we can reasonably assume that the thermal mass of the contaminant is much smaller than the thermal mass of the water and that compensating for the temperature variations does not remove the contamination events.

Abbate et al11 propose a model for the dependency of the water refractive index on temperature. The model is richer than we need in this context, thus we truncate the model at the first linear terms. With $n(T)$ the water refractive index, T the temperature, T_0 a reference temperature and B a given constant, we obtain:

$$n(T) \cong n(T_0) + B \times (T - T_0). \tag{4}$$

They estimate the factor B on basis of measurements and derived it as a general constant for large temperature variations. However, we are interested in a much smaller temperature scale and we also need to take into account the temperature dependency of the sensor itself. We propose to nevertheless apply the model in (4), but to locally estimate B and remove the temperature variations from the water refractive index. With this aim, we select a window of data of size W and for each time instant k we collect W samples of the water refractive index to build the vector

$$\mathbf{n}(k) = [n(k-W+1) \quad n(k-W+2) \quad \cdots \quad n(k)]^T \tag{5}$$

where $[\ ]^T$ indicates the transpose. With a similar procedure we define the temperature vector as

$$\mathbf{T}(k) = [T(k-W+1) \quad T(k-W+2) \quad \cdots \quad T(k)]^T . \tag{6}$$

We then apply a conventional linear regression approach to estimate the linear factor as

$$\hat{B}(k) = \frac{\mathbf{n}^T(k)\widetilde{\mathbf{T}}(k)}{\widetilde{\mathbf{T}}^T(k)\widetilde{\mathbf{T}}(k)}, \tag{7}$$

where $\widetilde{\mathbf{T}}(k)$ is the zero-mean version normalization of $\mathbf{T}(k)$ of (7). Equation (7) can be then used to obtain the temperature compensated water refractive index vector as

$$\widetilde{\mathbf{n}}(k) = \mathbf{n}(k) - \hat{B}(k)\mathbf{T}(k). \tag{8}$$

Equation (7) shows that the compensated water refractive index $\tilde{\mathbf{n}}(k)$ is obtained by subtracting a properly scaled version of the temperature, i.e. $\hat{B}(k)\mathbf{T}(k)$, from the original water refractive index. Event detection algorithms can be applied to the compensated water refractive index $\tilde{\mathbf{n}}(k)$.

As an example, we depict in Figure 8 the change in refractive index, the temperature as collected during an experiment and the resulting signal after temperature compensation. The change in refractive index Δn is composed of a default index difference Δn_0 between sensing and reference branch and varying term Δn_{water} that represents the change in water refractive index: $\Delta n = \Delta n_0 + \Delta n_{water}$. The correlation between phase and temperature can be clearly noticed. The resulting Δn of the temperature compensation algorithm shows 'white noise'.

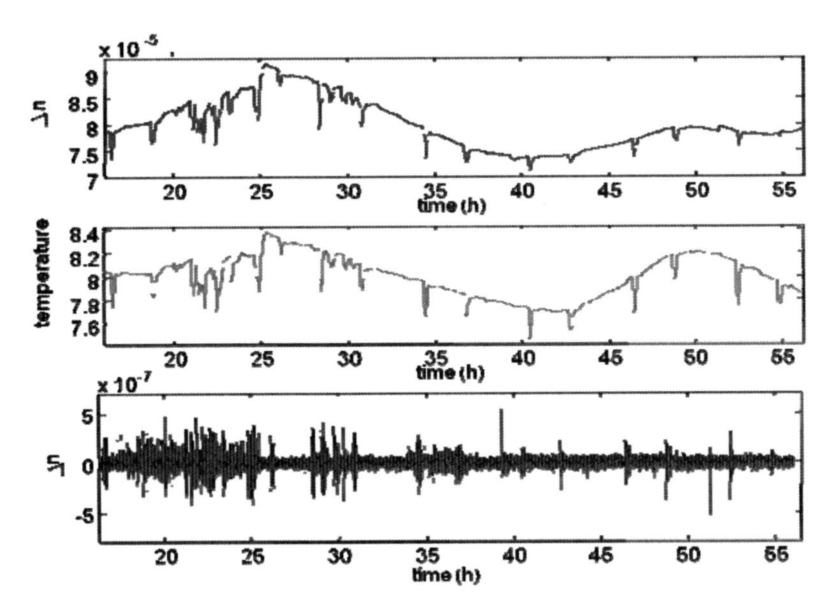

Figure 8 *Variations in the change in refractive index Δn and temperature as a function of time. Note that the scale in y-axis is reduced by two orders of magnitude.*

5.3 Event Detection Algorithm

We propose to use an event detection algorithm similar to the one presented by Smith[12], the so called edge detector. The basic idea is to compare the average value of the sensor output estimated in two separated windows. The difference between the two averages is the decision variable that is then compared to a threshold. The basic idea is illustrated in Figure 9 where we indicate with *B* the backward window, with *F* the forward window and with *D* a delay window. The averaged value in the backward and forward windows are indicated with m_B and m_F, respectively. The variable *S* is the standard deviation.

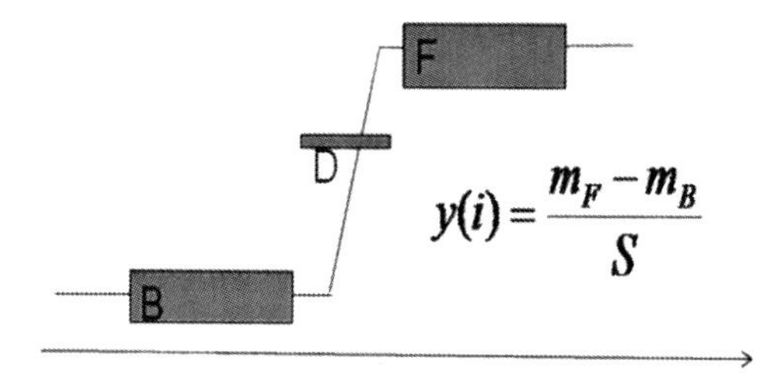

Figure 9 *Graphic description of the proposed edge detector*

The output of the event detector is the decision variable $y(i)$. We estimate the averages as

$$m_{B,F} = \frac{1}{\# B,F} \sum_{k \in B,F} x(k) \tag{9}$$

and the variance as

$$S^2{}_{B,F} = \frac{1}{\# B,F - 1} \sum_{k \in B,F} \left(x(k) - m_{B,F}\right)^2, \tag{10}$$

where the symbol # indicates the cardinality of the set, i.e. the size in sample of the windows. The decision variable is then

$$y(i) = \frac{m_F - m_B}{S}. \tag{11}$$

The approach proposed by Smith[12] suggests calculating the variance S^2 in such a way to limit the impact of the edge size on the estimation of the variance itself. We do not directly apply the solution presented by Smith. Actually, we propose to apply equation (9)-(11) twice on the same set of data. This means that the output $y(i)$ of (11) is used as the input $x(k)$ of (9) and (10). The reason for this choice is related to the characteristics of the background noise. In the data collected via experiments, there were some cases in which the variations in the background water refractive index were quite sharp. If an event happens during these sharp variations, the detection algorithm could not detect it. By applying twice (9)-(11), we could identify also those cases, thus improving the performance of the detection algorithms.

5.4 Simulation Results

We consider the water refractive index collected by the sensor in a real environment to evaluate the performance of the event detection algorithm. The data were collected by monitoring the tap water in an office building in the Netherlands. The events were added via software. The shape of the events has been determined by the spiking experiments with the contaminants, see chapter 4, and the amplitude has been properly normalized to a DAL level. The performance is showed in terms of receiver operator characteristic (ROC) curves that plot the true positive rate (TP) versus the false positive rate (FP)[12]. The ROC curve shows how the trade-off between the true positives and the false positives changes depending on the threshold. Each point of the ROC curve is obtained for a fixed threshold.

For instance, assume that the decision variable is strictly larger than zero and that there is no noise. Then, if we set the threshold value to zero, the algorithm must detect all events and it must return a true positive rate of 1. For the same threshold value, the false positive rate is also equal to 1 because all false positives are detected. By increasing the threshold the algorithm will return a different balance between true positive and false positives.

We consider the event obtained with aldicarb scaled down to a concentration of 3.6 mg/L, which corresponds to a DAL-2 level for adults, see Figure1. The algorithm does not need to know which contaminant generated the event; it is designed to detect abrupt changes. The amplitude of these changes is determined by the experimental results. The ROCs are obtained by software simulations which randomly select a chunk of sensor data, sum up an event, run the algorithm and collect the true positives and false positives. This procedure is iterated many times (about 10000 times). Figure 10 shows a zoomed version of the ROC curves for aldicarb DAL-2 when the edge detector is applied once and twice. The performance is good: For a 1% of false positive all methods provide more than 98% of true positives. We can also observe the effect of the temperature compensation algorithm as we switch it ON or OFF. It appears that the temperature compensation provides a significant improvement. It also appears that the temperature compensation boost the performance of the single edge detector to the same performance as the double edge detector.

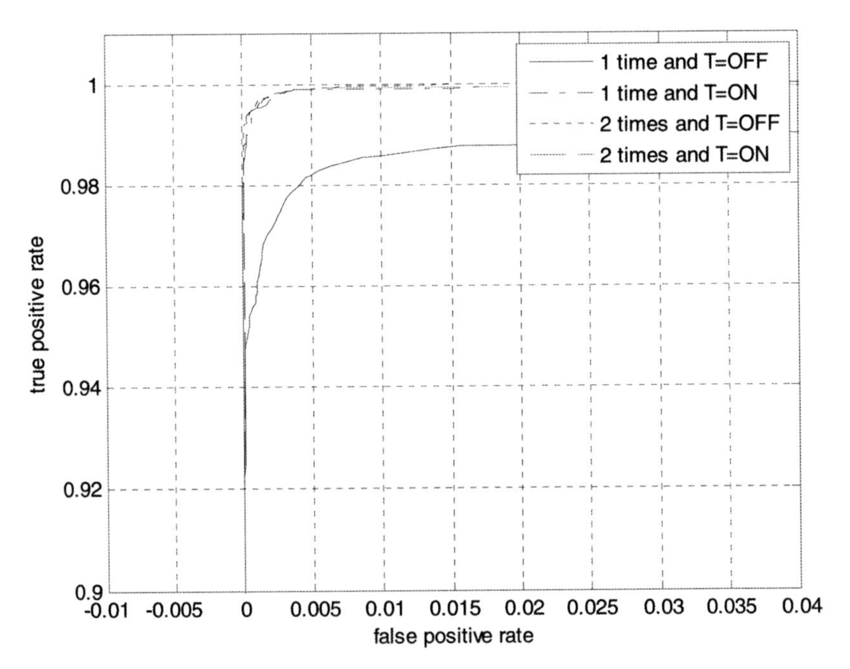

Figure 10 *ROC curve, aldicarb DAL-2 for adults (3.6 mg/L). B=F=20. D=5. The units of B,F and D are in samples and each sample corresponds to 1s. Four cases are shown: 1 time T:OFF, corresponds to one application of the edge detector with the temperature compensation algorithm switched OFF.*

To better appreciate the differences between the application of the edge detector once or twice, we run the same simulations but with a much lower concentration of 0.6 mg/L corresponding to a DAL-2 for an infant. This means that the events added to the simulation are six times smaller than the one used in Figure 10. The results are shown in Figure 11 and they are worse than those in Figure 10, as expected.

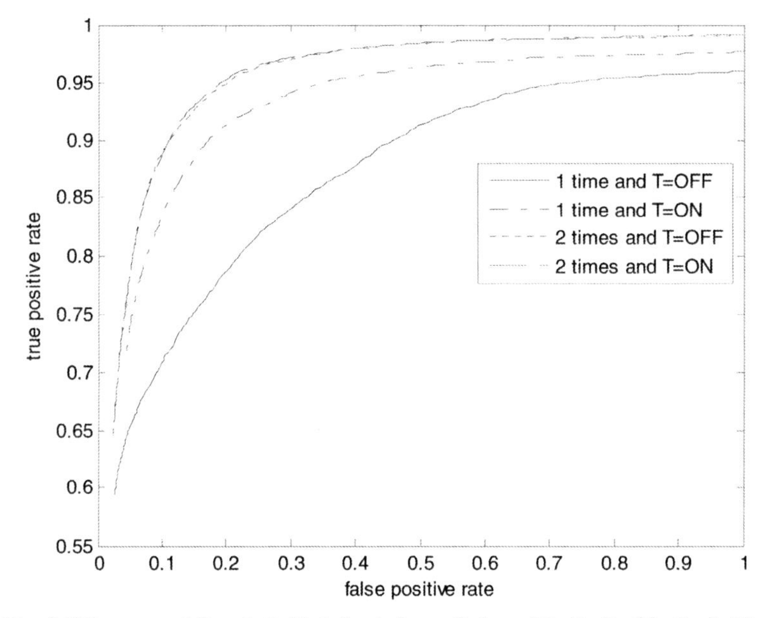

Figure 11 *ROC curve, aldicarb DAL-2 for infants (0.3 mg/L). B=F=20. D=5. The units of B,F and D are in samples and each sample corresponds to 1s. Four cases are shown: 1 time T:OFF, corresponds to one application of the edge detector with the temperature compensation algorithm switched OFF.*

However, in this case we can clearly appreciate the improvement of applying the edge detector twice while in Figure 10 it seems that the temperature compensation and the application of the edge detector twice provided the same improvement. We can also notice the different scale and the reduced performance due to the much smaller events.

6 CONCLUSIONS

Based on the template of Acute Exposure Guidelines Levels it is feasible to derive short time exposure levels (Drinking water Alert Levels) for drinking water contaminants. In accordance with the AEGLs three severity categories for public health effects can be distinguished (DAL-1, DAL-2 and DAL-3). DAL values can be derived for adults and infants separately, with reference to the differences in drinking water uptake on a body weight basis.

In order to evaluate the short time exposure to drinking water contaminants the Drinking water Alert Levels are useful. Comparison of these alert levels with the detection limit of Optiqua's Mach Zehnder interferometer shows that for the selected substances detection is possible in a concentration range where no severe health effects are expected.

Temperature variations have a large impact on the sensor response. With a model of temperature dependence of the refractive index an algorithm for temperature compensation has been applied.

Application of the edge detecting algorithm on time series of natural drinking water quality, superimposed with spiking signals, shows that contamination events can be detected. These results suggest that the Optiqua sensor can be used as an early warning monitor in a drinking water network to detect harmful contaminants.

7 FURTHER RESEARCH

The promising results of the Aqua Vitaal project are now followed up with further development and testing. There are four areas of further research that are now actively worked on:

- Event simulation in drinking water instead of Milli-Q water

Vitens is working together with Optiqua to simulate contamination events in a continuous flow test system using regular drinking water as running medium. In addition to the Optiqua sensor, the test system contains a spectrum of other sensors to monitor regular process parameters (pH, conductivity). Vitens and Optiqua are deriving detection limits for a wide range of substances, and are validating and refining calibration factors and detection limits in real drinking water for the substances investigated in this paper.

- Analysis of normal water quality fluctuation in a real distribution network

The fundamental driver for practical effectiveness of an early warning system is the ability to discriminate the signal of an event from the background signal. Vitens and PUB in Singapore are currently collecting continuous data on various strategic locations in the network, using the Optiqua sensor. Long time series of continuous data at different locations will provide the opportunity to validate the robustness of the sensor in different water sources against a background of natural variation including seasonal patterns. Also the effect of a slowly varying contamination will be investigated

- Duration testing in a real distribution network

Optiqua will use the results from the testing at the Vitens and PUB locations to evaluate the durability and long-term performance of the sensor.

- Analysis of locations for sensor placement

An important area of research is into the question on where to locate sensors within the distribution grid. A dense network of sensors in the distribution grid would have the benefit of detection of a contamination close to the source. Contamination of a relatively large area above the DAL values would mean that at the source the concentrations are much higher due to the effect of mixing and diluting. The location of the sensor is likely to have an important impact on the performance of the sensor. Best locations for optimal benefits will depend on characteristics of the network (e.g. number of crossing pipes, flow rate, numbers of end-points).

Acknowledgements
The authors would like to thank Ton Koster for his contribution.

References

1 Guidelines for drinking water quality, third edition, incorporating first and second addenda, World Health Organisation, Geneva, Switzerland, 2008
2 National Research Council, Standing Operating Procedures for Developing Acute Exposure Guideline Levels for Hazardous Chemicals, National Academy press, Washington DC, USA, 2001

3 R.Solecki, L.Davies, V.Dellarco, I.Dewhurst, M.van Raaij, A.Tritscher, Guidance on setting of acute reference dose (ARfD) for pesticides, *Food Chemical Toxicology*, 2005, **43**, 1569
4 Toxicology and Hazard Assessment Group, Standing Operating Procedures for the Development of Provisional Advisory Levels (PALs) for Chemicals, Oak Ridge National Laboratory, Oak Ridge TN, USA, 2008
5 A.V. Wolf, M.G. Brown, and P.G. Prentiss, Concentration properties of aqueous solutions, Handbook of Chemistry and Physics, CRC Press, 65th edition, 1984-1985, D-222
6 J.J. Ramsden, A sum parameter sensor for water quality, *Wat. Res.* 1999, **33**, No. 5, 1147
7 R.G. Heideman, P.V. Lambeck, Remote optical-chemical sensing with extreme sensitivity: design, fabrication and performance of a pigtailed integrated optical phase-modulated Mach-Zehnder interferometer system, *Sensors and Actuators,* 1999, **B 61,** 100
8 P.V. Lambeck, Integrated optical sensors for the chemical domain, *Measurement Science and Technology*. 2007, **17,** R93
9 R. Murray, T. Haxton, S.A. McKenna, D.B. Hart, K. Klise, M. Koch, E.D. Vugrin, S. Martin, M. Wilson, V. Cruz, and L. Cutler, Water Quality Event Detection Systems for Drinking Water Contamination Warning Systems- Development, Testing, and Application of CANARY, available at http://www.epa.gov/nhsrc/pubs/600r10036.pdf
10 N. C. Gallager, G.L.Wise, A theoretical analysis of the properties of the median filters, *IEEE Trans. On acoustics, speech, and signal processing*, 1981, ASSP-29, **6**
11 G. Abbate, U.Bernini, E. Ragozzino, and F.Somma, The temperature dependence of the refractive index of water, *J. Phys. D: Appl. Phys.*, 1978, **11**
12 D.A. Smith, A quantitative method for detection of edges in noisy time-series, *Philos. Trans. R. Soc. Lond. Biol Sci.*, 1998, **353**, 1969

DETECTION AND IDENTIFICATION OF MICROBIAL CONTAMINATION

R. Aitchison[1], C. Heller, U. Reidt, A. Helwig and A. Friedberger[2]

[1] CBRNE Advisor, Cassidian (an EADS company), Quadrant House, Celtic Springs, Coedkernew, Newport , NP10 8FZ, UK
[2] EADS Innovation Works, IW-SI – Sensors, Electronics & Systems Integration, 81663 Munich, Germany

1 INTRODUCTION

Contamination of water supplies is an inevitability. Some natural causes may have a degree of predictability, in as much as the track of a storm and its likely impact can be anticipated (with variable success). The utility of any early warning is however limited and very few incidents can be predicted with meaningful confidence.

Existing monitoring technologies potentially afford sufficiently rapid detection and identification of common natural physical and chemical contaminants to allow timely action to prevent or limit the extent of spread and to permit prompt initiation of appropriate remedial action. A notable exception is reliable detection and identification of microbial contamination.

This paper explores some of the challenges associated with the control of bio-contamination and describes an emerging technology that promises to bring timely detection and identification of microbial contaminants within the reach of the water supply industry.

2 DISCOVERY

Contamination can occur at any point in the water distribution system from source to consumer but the worst place to discover it is at the point of consumption. By this stage it is too late to avoid any potential harm to the consumer but without the ability to detect contamination upstream, the consumer becomes the first indicator of potential problems.

Fortunately, the water treatment and distribution infrastructures of the world's developed economies ensure that such incidents are extremely rare. Nevertheless, the potential remains for transmission of pathogens to the population due to possible contamination at various points in the water supply system.

'Natural' contamination events may be the most common, but the risk to the general public of contracting a serious illness as a result of either natural or accidental contamination of water with pathogens is very low. Conversely, the risk of any deliberate contamination is low but the potential for harm to affected consumers may be enormous.

Taken together, there is a strong argument for monitoring water supplies for specific pathogens at various points in the system.

Whether or not the public health and economic arguments favour widespread monitoring, current technology does not allow sufficiently rapid and reliable microbial detection / identification for such routine monitoring to be worthwhile. Historically, the only reliable means of identifying dangerous microorganisms is the use of techniques based on culturing methodologies that take a minimum of around 2 days to produce definitive results. New technologies are now beginning to emerge that promise much more rapid identification and quantitation of both pathogens and indicator organisms such as *E. coli*. By combining developments in miniaturisation with novel filter material and process automation we have demonstrated that continuous, close to real time monitoring may be becoming a practical proposition.

For protection against bio-terrorism, the standard detection thresholds (e.g. 100 colony forming units (cfu) /ml or 1 cfu/100ml, resp. for *E. coli*) may need to be improved upon and identification techniques such as PCR (Polymerase Chain Reaction) will be required. Where significantly higher detection sensitivities are required, they are likely to be achieved through a combination of analyser refinement and the application of enhanced sample collection and preparation techniques.

3 RAISING THE ALARM

Where routine monitoring is unable to identify a problem in less than 24-48 h, the most likely trigger for an alarm is mysterious medical symptoms among consumer communities. If the distribution network is protected by in-line detectors providing readings on an hourly (or less) basis, there is every prospect of being able to limit the potential harm to consumers considerably. However, this would require a rapid reaction to alerts generated by the detectors and even a relatively low rate of false positive alarms would have serious ramifications for both the water company and its customers. Detection equipment that simply monitors overall microbial load is therefore not sufficient. A reliable automated identification would solve this problem and make automated alert and water network control a practical proposition.

Almost as important as the reliability of the identification is the alert raised, the information provided to water network operators and the speed and appropriateness of action initiated automatically. To achieve this relies as much upon the communications and information / control systems as the monitoring equipment itself.

4 BALANCING ACT

While infection of consumers with disease causing microorganisms has been largely eradicated in the developed world, the treatments used to give this assurance are not inexpensive and present their own potential health risks. For example, the higher the dose of anti-microbial agent used to control bio- contamination, the greater the potential health risks from either the agents themselves or harmful by-products of the process. Although secondary, there is also the consideration of the financial and environmental consequences of using unnecessarily high quantities of chemicals that can be both expensive and consume large amounts of energy in their manufacture and distribution.

Establishing and maintaining the optimum balance demands reliable, continuous monitoring of both upstream and downstream water to ensure:

a) dosing is varied to accommodate fluctuations in the microbial profile and loading of the water
b) treatment has been effective.

The automated microbial analyser technology presented here could be refined to construct a control system that would tailor precisely the treatment regime to maintain both an appropriate safety margin and the minimum consumption of biocides.

In this instance, the detectors and effectors are likely to be located in the same premises or at least on the same site but for the protection of whole water supply networks, detector networks are likely to be required. To achieve an appropriate balance between confidence in protection of consumers and avoidance of false alarms, sophisticated expert system capabilities can be employed to improve both accuracy of identification and tracking of contamination and to model its spread through the system. By applying similar modelling to an appropriate distribution control infrastructure, contaminated water flows can be isolated or diverted and alternative supplies brought on-stream as necessary.

5 DETECTION TECHNOLOGY

Improved microbial detection technology is required to:

- Provide early contamination alerts
- Allow timely isolation / diversion of contaminated supplies
- Allow optimised dosing of anti-microbial treatments
- Monitor efficacy of remedial action following contamination incidents

All of these applications demand automated monitoring systems providing continuously high sampling frequency and high detection sensitivity for targeted microorganisms.

A rapid analytical technique has been enhanced by researchers to create instruments potentially capable of meeting such requirements, initially for bacteria but potentially also for eukaryotes, fungi, viruses and biological molecules.

The primary constituent elements of the technology are silicon wafer based microporous membranes, automated high specificity analytical techniques with high sensitivity optical detection systems and process controls to cleanse the equipment in preparation for the next analytical cycle. Figure 1 shows a schematic of the basic architecture of the microbial analyser.

One of the most critical elements of the design is the microporous membrane. Its key characteristics are its dense pattern of highly regular perforations of extremely uniform pore size (see Figure 2) and its non-stick properties that make effective removal of captured material a practical proposition.

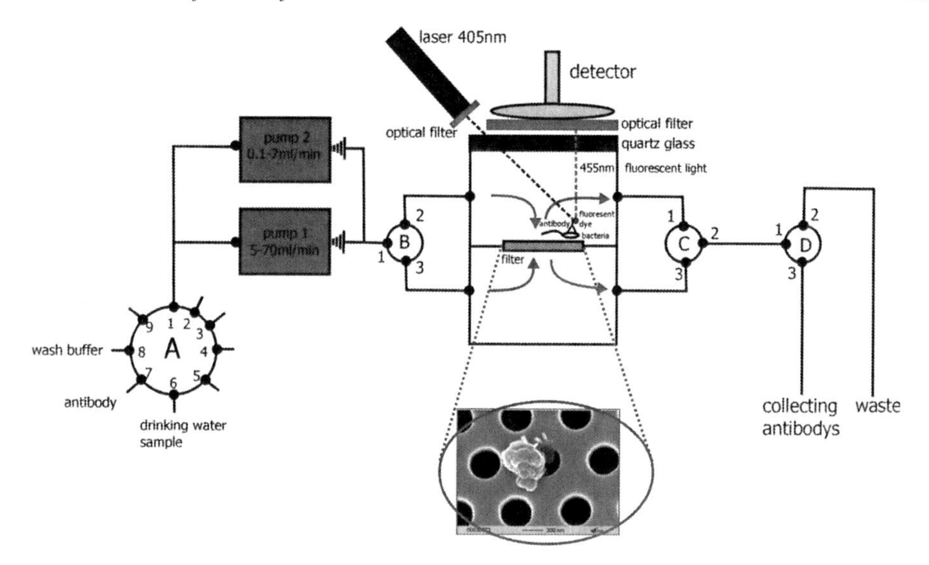

Figure 1 *Schematic of the main elements of the microbial analyser design*

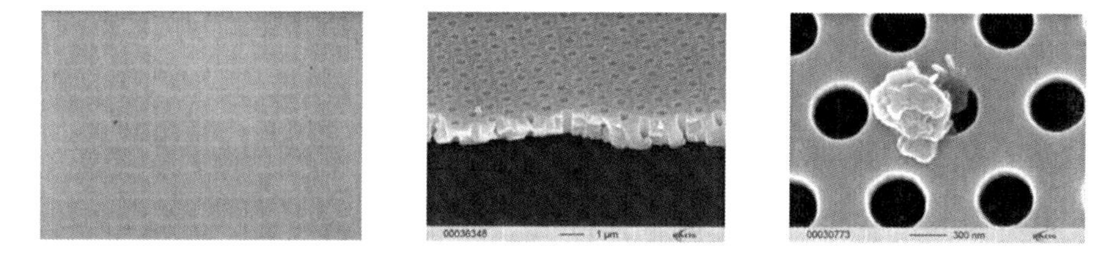

Figure 2 *25mm² silicon chip perforated with 450nm pores and showing material captured on its surface.*

The design optimises flow rate and capture of the targeted microorganisms, retaining them on its surface and allowing them to be removed readily by back-flushing. The photographs below (Figure 3) of the membrane before and after flushing indicate how well this works[1].

The bacteria trapped on the filter surface can subsequently be subject to bioanalytical techniques such as immunoassays or PCR to optimise detection and identification of targeted microorganisms.

For simple intercalating dye based techniques the instrument's operation can be summarised as shown in Figure 4.

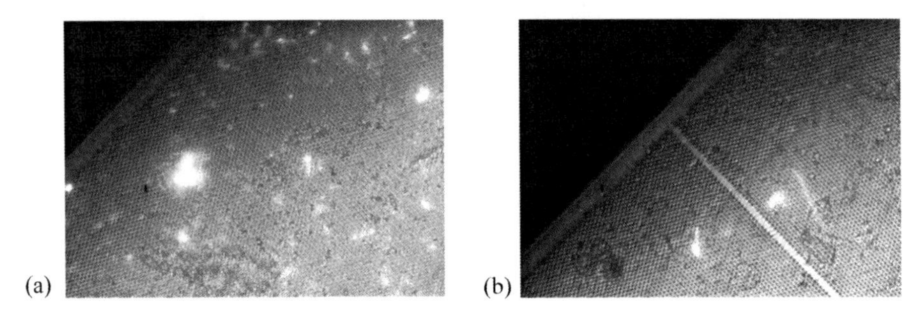

Figure 3 *Filter surface following capture of bacteria in a 40ml sample containing 5,000 cells/ml before (a) and after (b) washing with deionised water. Removal of cells was verified by cultivation.*

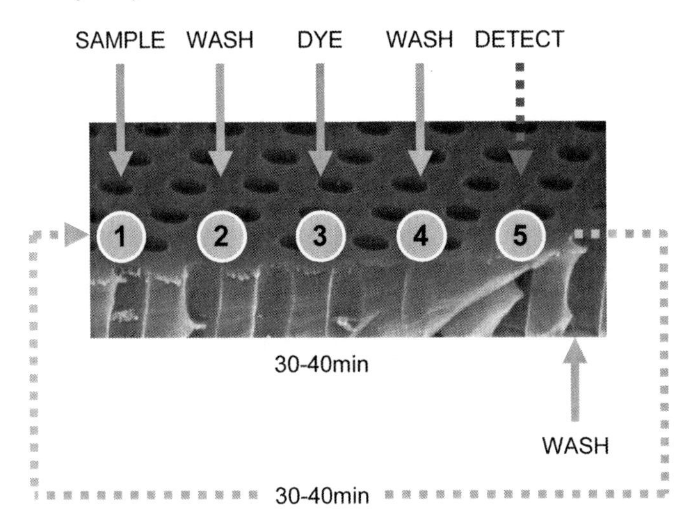

Figure 4 *Indicative sequence of operations performed by the microbial analyser to complete a typical analytical cycle employing detection by antibody conjugated fluorescent dye.*

6 SECURING MONITORING AND CONTROL SYSTEMS AGAINST CYBER ATTACK

Continuous monitoring and rapid action to isolate contaminated supplies clearly has major benefits for consumers and water companies alike. The sophisticated Supervisory Control and Data Acquisition (SCADA) systems necessary to ensure prompt and appropriate response to alert signals from monitoring devices are however potentially a source of vulnerability. The possibility of critical national infrastructures such as water and power being disrupted by cyber-attacks is increasingly being recognised as a serious potential threat. Although contaminating water supplies with dangerous pathogens may remain the greater prize for terrorists, the ability to sabotage treatment plants or cause water supplies to be shut down may be considered almost as effective, yet simpler and lower risk.

Highly effective methods are available to counter this threat by securing SCADA systems against cyber-attack. All potential modes of attack can be identified and the level of vulnerability determined. Protection can then be tailored to the specific requirement but options, which can be used individually or in combination, include:

- Communications link encryption (for connections between detectors, control centre and effectors)
- Network based encryption (to secure all SCADA traffic across shared network infrastructures)
- Boundary switch / firewall protection (to prevent access to the 'private' network through connections to 'public' network services such as the internet)
- Access control systems (to prevent access to the SCADA system by unauthorised personnel)
- Malicious code detection / quarantine and disinfection systems (to prevent viruses, worms, trojans, etc. from infecting SCADA systems or the platforms on which they are hosted).

By modelling the impact of a cyber-attack, the application of some or all of these techniques can be tailored to establish the appropriate level of end-to-end or domain-bounded security. This approach ensures that the necessary level of protection is achieved in the most cost-effective manner. For example, if an encryption system and powerful access control features are used to essentially isolate a SCADA system, it may not be necessary to protect the shared communications infrastructure over which it operates. Also, if alerts from individual monitoring devices are not in themselves sufficient to trigger precipitous action, there may be no real necessity to protect the communications link if the control system is secured properly against unauthorised access.

7 CONCLUSIONS

Historically, the ability to develop and implement effective monitoring and control systems for microbial contamination has always been limited by the inability to detect and identify microorganisms sufficiently quickly or reliably.

- The technology presented here promises to remove this limitation by enabling instrumentation to be manufactured that allows:
- Automated continuous in-line monitoring
- Detection and identification of specific pathogens and indicator organisms
- Low limit of detection
- Minimised false positive alerts
- Linking to water supply control systems for rapid containment of contamination

Microbial analysers employing this technology could be sited at any strategic locations in the water supply system, including:

- At point of abstraction
- Before and after treatment
- At service reservoirs
- At major distribution network nodes
- On entry to sensitive sites of consumption (e.g. hospitals).

This kind of instrumentation would enable monitoring to:
- Provide contamination alerts and prompt initiation of remedial work
- Allow consumers to be warned in advance of exposure to contaminants
- Allow timely isolation / diversion of contaminated supplies
- Allow optimised dosing of anti-microbial treatments
- Assess the efficacy of remedial action following contamination incidents

The SCADA system employing continuous microbial monitoring can be protected from cyber-attack by:
- Implementing secure communications between sensor, control room and effectors
- Isolating the SCADA system from corporate information infrastructures
- Securing corporate information infrastructures against external intrusion
- Detecting and neutralising malicious code
- Implementing robust identity management systems to prevent unauthorised access

Acknowledgements
A significant part of the overall system has been accomplished in cooperation within the BMBF funded project OptoZell (funding agency: VDI Düsseldorf). We gratefully acknowledge funding through the BMBF and contributions from the project partners L. Meixner and K. Neumeier (Fraunhofer-Institut für Zuverlässigkeit und Mikrointegration), P. Lindner, R. Molz and H. Wolf (Universität Regensburg, Institut für Medizinische Mikrobiologie und Hygiene), N. Zullei-Seibert (Westfälische Wasser- und Umweltanalytik) and G. Preuß (Institut für Wasserforschung).

Produced on behalf of Cassidian
CASSIDIAN, an EADS Company, is a worldwide leader in Global security solutions and systems, providing lead system integration and value-added products and services to civil and military customers around the globe: air systems (aircrafts and Unmanned Aerial systems), land & joint and naval systems, intelligence and surveillance, cybersecurity, secure communications, test systems, missiles, services and support solutions.
CASSIDIAN, "Defending world security". www.cassidian.com

For further information, please contact robert.aitchison@cassidian.com

References

1. Reidt, U., Chauhan, L., Müller, G., Molz, R., Lindner, P., Wolf, H. and Friedberger, A., Reproducible Filtration of Bacteria with Micromechanical Filters, Journal of Rapid Methods & Automation in Microbiology, 2008, **16** (4), 337–350

VALIDATION OF A WATER QUALITY MONITORING PLATFORM IN BARCELONA DRINKING WATER TREATMENT PLANT

R. López-Roldán[1], S. González[1], J. Ribó[2], J. Appels[3] and J.L. Cortina[1]

[1]CETaqua, Diagonal 211, Barcelona, E-08018, Spain
[2]Centre for Research and Innovation in Toxicology (CRIT-UPC). Technical University of Catalonia, Campus Terrassa, zone IPCT. Ctra Nac.150 km 14,5 08227 Terrassa, Spain
[3]microLAN b.v., Biesbosweg 2, NL-5145 PZ Waalwijk, Netherlands

1 INTRODUCTION

Water bodies quality monitoring is a must nowadays, but not only for the preservation of the ecological status. Final uses of water, especially for irrigation, industrial application or potable purposes, demand a good quality. Treatment technologies are implemented when quality does not fit requirements for these specific uses. Performance of these technologies and quality of produced water are also very dependent on the characteristics of the resource at the intake.

Actions have also been implemented to improve the quality of these natural bodies. Impact of these preventive and/or corrective actions should be monitored to follow-up the efficacy and to meet normative standards.

At the European level, quality fulfilments to be met are mainly addressed by the Water Framework Directive (WFD, 2000/60/EC)[1] and the Priority Substances Directive (2008/105/EC)[2] where Environmental Quality Standards (EQS), that is concentration, and Annual Average (AA), that is total loads, are established for a list of 33 pollutants. Performing laboratory analysis of these 33 pollutants implies a large number of analyses and it should be taken into account that it is not enough to measure whether or not the priority substances concentration is below the EQS or not because this is not necessarily representative of the water status. Moreover, spot sampling campaigns, the most common approach for analysing these compounds are costly and labour-intensive and not sufficient to have an accurate picture of the chemical and biological status of water quality.
For all of these reasons, new tools should be established to get all the information needed.

One of the most sensitive uses of natural waters is their treatment for drinking purposes. Source water need to comply with the requirements that make it suitable for their treatment in the drinking water plant facilities. European and national legislation has also been established to protect health of drinking water consumers. Drinking Water Directive (98/83/EC)[3] and recent application of Water Safety Plans (WHO Guidelines[4] or ISO 22000[5]) imposes strict regulation on the monitoring of water quality indicators, mainly addressed to protect human health. Normative plus technology requirements make it necessary to monitor water at the intake of treatment plants.

2 LOCATION / CASE STUDY

The Barcelona area (NE of Spain) is being supplied with drinking water originating from different sources. Water scarcity situations, typical in the Mediterranean area, require a complex management taking into account availability of different water sources: - Llobregat and Ter rivers, groundwater, sea and brackish water, all combined, in different blends, according to the season of the year and the exact location in the Barcelona area.

Although having this complex system of water supply, the main source is surface water coming from the Llobregat River. Previous studies and monitoring campaigns show the presence of a significant number of families of pollutants[6-8]. Some physico-chemical parameters are continuously monitored. The values of these parameters exhibit significant variations, especially due to the seasonal effects. Figure 1 shows information on the flow variation at the intake of Sant Joan Despí Drinking Water Treatment Plant (DWTP). Uneven distribution of rainfalls during the year provokes sudden alterations of river characteristics, especially flow and turbidity due to runoff in heavy rain events.

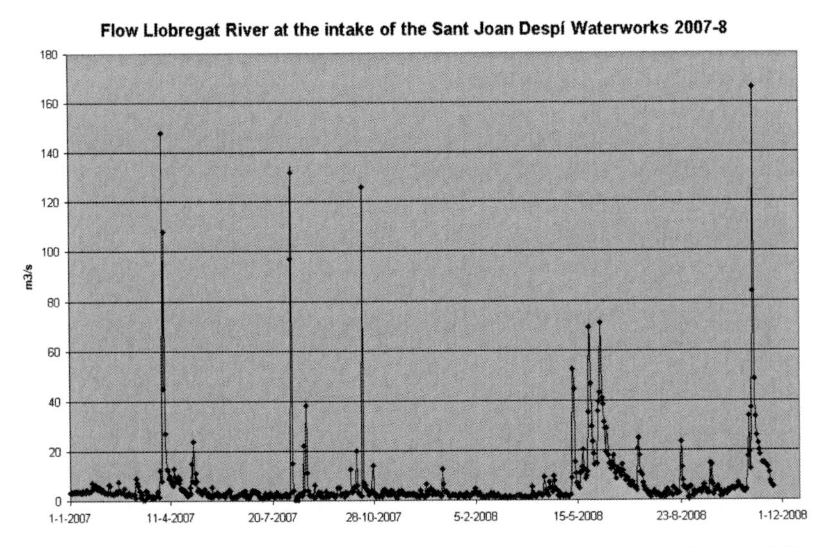

Figure 1 *Seasonal flow variation of Llobregat River at the Sant Joan Despí DWTP*

A sudden alteration of quality parameters could be interpreted as an episode of accidental spill or illegal discharge into the river, not only impacting ecosystems but creating a situation where polluted water could enter the water treatment plant potentially affecting the quality of drinking water. If this situation occurs, intake from the relevant surface water is closed and groundwater is pumped from the aquifers in the area.

3 VIECO PROJECT

In order to successfully achieve this objective a monitoring platform hadv to be designed. The platform integrates off-line and on-line measurements and in-line integrating sampling tools in order to obtain detection of a wide range of different parameters and to have integrated responses according to water quality.

In a first step, the project was directed to the integration of these new sensing technologies (including chemical and bio-sensors) with the existing ones, in order to obtain automatic and autonomous detection of different parameters and to have integrated responses according to water quality. In a second step the monitored data are being integrated to establish a relationship between field ecological observations and measured parameters related to chemistry, toxicity, etc and additionally to distinguish the contribution of unknown (not measured) causes. The platform will provide a low-cost management tool for providing different levels of environmental information in surface waters, with a high degree of integration of data processing and interpretation of whole results.

3.1 Objectives

The main objective of this project is to contribute to the exhaustive characterization and further enhancement of water quality by developing and validating a cost-effective monitoring platform to provide environmental indicators of ecological, chemical and biological status of surface waters. Validation will be performed at the main DWTP in Barcelona.

This objective will be achieved by providing the platform with emergent tools and validating results with traditional methods. Environmental indicators will be offered to operators and Government relating to water ecological status integrating data provided by the tools that have been validated, contributing this way to implement the WFD.

3.2 Actions and means involved

In order to reach the above mentioned objective, the following sets of actions are being carried out:

3.2.1 Definition of the off-line chemical sensing platform

Analytical methods have been optimized for the analysis of a selection of priority and "emerging" compounds. Liquid chromatography coupled to tandem mass spectrometry has been the technique selected. Different methods for detection and quantification were developed for pesticides, pharmaceuticals and fullerenes. A list of contaminants has been selected on the basis of: a) their high use and/or production, b) their significant aquatic toxicity, c) their fate in the aquatic environment (high stability and low biodegradability), d) their poor removal during activated sludge wastewater treatment.

3.2.2 Integration of sensors for the on-line bio and chemical sensing platform

In this task on-line chemical and biological sensors are currently being validated for their integration on the monitoring network stations.

For that purpose, a biological toxicity monitor using luminescent bacteria (Toxcontrol, microLAN, Netherlands) coupled to a diode array UV-VIS spectrophotometer probe is being tested (spectro::lyser, s::can, Austria). On-line measurement of several parameters (TSS, COD, TOC, NO3-) using a single UV-VIS probe could help to establish relationships among these parameters and characterize surface water before entering the plant. Toxicity is being measured by the decrease of the luminescence of *Vibrio fisheri* bacteria and an alert signal being generated. A Solid Phase Extraction (SPE) concentrator prototype is also being tested to increase sensitivity of this biomonitor to toxic substances. This combination is being assessed for pollutants more commonly found in the studied area and those posing major risk to water production including pesticides, personal care products, surfactants, etc.

Combination of complementary techniques will allow for the verification of alarm signals reducing false alarm rates. Where a chemical analytical monitoring system identifies and quantifies specific water contaminants, biomonitoring gives an indication of the total quality, including the effects of unknown toxic substances.

An equipment for indirect on-line measure of the Biological Oxygen Demand (DBO Optosen 50, Interlab, Spain) based on the measurement of oxygen consumed by bacteria immobilized on a plate, will be validated not only at the intake of the DWTP but at the effluent outlet of a Wastewater Treatment Plant (WWTP).

3.2.3 Integration of in-line integrating sensing tools

Methodologies based on passive samplers have been developed for in-line integrating monitoring of water quality as they provide more realistic results than conventional methods. Passive samplers are introduced into surface water streams and they adsorb certain chemicals during a period of time (normally between 2 and 4 weeks), commonly though a diffusion-limiting membrane. Afterwards the samplers are removed and brought to the laboratory, where target compounds are extracted and quantified. As a result, these devices calculate a time weighted-averaged concentration (TWA), which is more representative than results of conventional methods using spot samples, because they mimic how these pollutants are taken by animals and plants and how they bioaccumulate. In addition to this, as they integrate concentration over a period of time, they provide an average, not a snap shot of the situation when a sample is taken.

The diffusive gradient in thin film (DGT) devices was tested for the analysis of metals. Polar Organic Chemical Integrative Samplers (POCIS) were tested for the analysis of polar compounds like pharmaceuticals, pesticides and surfactants, while an automatic device for constant flow integrative analysis (CFIS) was developed and tested for non-polar organic compounds like PAHs, PCBs, pesticides, phthalates and surfactants.

3.2.4 Implementation of the biomonitoring platform for assessment of pollution induced effects on biofilm communities

A bio-monitoring platform has been developed using field diagnostic tools for estimation of water quality. The aim of this task has been directed to the implementation of tools for community based toxicity assessment of site specific toxicants by integrating structural, physiological and functional parameters of microbenthic communities and community based ecotoxicological assessment approaches. Changes in community structure (biomass, taxonomy and metabolic profiling) have been studied under different toxicity exposures. Effects on functional parameters from biofilms will be assessed through a variety of methods.

3.2.5 Risk evaluation of toxicity in aquatic ecosystems

The proposed task will be focused on the integration, at river basin scale, of several parameters measured by sensors, classical analytical chemistry methods and bioassays to define indicators to assess the impact and the effects of ecological field observations. Thanks to this methodology, it will be possible to relate the field ecological observations (e.g. abundance of species) to parameters related to habitat, chemistry, toxicity, effluent and to distinguish the contribution of unknown (not measured) causes. The correlation between the monitored and quantified parameters within the project versus the observed ecological parameters will be carried out using generalised linear models (GLM) and a flexible quadratic model will finally distinguish the influence of each group of variables on the final observed value and, thus, attribute the effects to parameters.

3.3 Results on validation of on-line sensors

3.3.1 Spectrophotometer probe

The diode array UV-VIS spectrophotometer probe was tested, in a first stage of the project, in the surface water entering the DWTP. Measures were programmed automatically every 10 minutes. The probe is connected to an air compressor for cleaning the window every 5 measurements. The profile of the spectrum from 200 to 750 nm is recorded during each measurement. The calibration model shows values from pre-established parameters with the information obtained from the spectral profile. In this case, on line measurements of several parameters: - Turbidity, Total Organic Carbon (TOC), nitrates and SAC-254 are performed. Correlations are established among values obtained from the probe and the analysis performed at the laboratory. This comparison will give us information on the performance of the instrument and it will be used by the probe itself for local calibration.

The main advantages of using a multiparametric probe are: - compilation of information about several parameters using a single technology resulting in a lower overall cost for purchase and maintenance; real time continuous monitoring of the water at the intake (potentially applied to other steps in the process of drinking water treatment), making it suitable to be used as an alert system; recording of the whole spectrum from 200 to 750 nm to give us the chance to add new parameters or compounds and extract concentration values from the footprint.

The main difficulty found when validating the probe was the unstable conditions of the Llobregat River. The water matrix changes fast in Mediterranean rivers (e.g. turbidity can change from 30 NTU to 1000 NTU in a few hours). The probe has been calibrated for a concentration range of parameters narrower than the range found in real conditions. That means that in certain episodes (heavy rains), data is out of range and cannot be recorded. The year the validation process took place (2010) had extraordinarily high rainfall so extreme results, especially turbidity values, occurred quite frequently. Figure 2 shows turbidity values from November 2009 to June 2010 (values above 300 NTU are not shown).

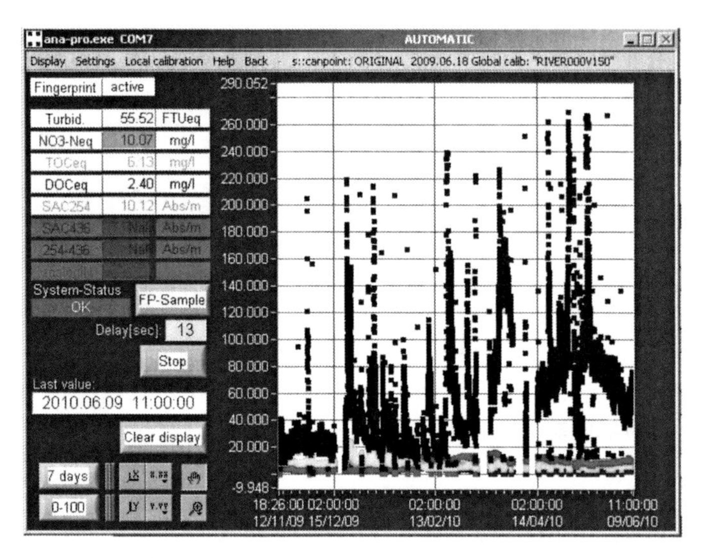

Figure 2 *Screen shot of spectrophotometer probe software showing variations in turbidity values*

3.3.2 Biological toxicity monitor

A biological toxicity monitor using luminescent bacteria is being tested. The equipment is completely automated. Measurements of the incoming flow of water are carried out every 30 minutes having a quasi real detection of the global toxicity of the compounds being present in the water. Toxicity is being measured by the decrease of the luminescence of *Vibrio fisheri* bacteria when being in contact with sample. Exposure time has been set in 15 minutes. An alert signal can be established at a certain value of light inhibition.

At the Llobregat River, it was found that the background levels of toxicity are typically in a range of ± 20% of light inhibition. As it was stated before, the water matrix changes due to weather conditions. When turbidity is high, we can see some positive toxicity due to the decrease of light arriving at the photomultiplier light detection system in the equipment (Figure 3). In low turbidity episodes, toxicity values are negative because of the high content of nutrients in Llobregat water that results in an increase of the metabolism of the luminescent bacteria and therefore, the emission of light.

In one of the experiments performed, toxicity monitor were used to analyse, at off line mode, 6 samples of surface water at the intake of DWTP plus other 6 samples of treated wastewater at Abrera WWTP, discharging to Llobregat river upstream the DWTP intake. A system was designed using a pump so the same sample flowed in a loop and it could be measured several times. Having different measures of the same water is giving us information on the repeatability of the method. Different aliquots of the same sample were analysed by a new prototype version of the toxicity monitor to compare performance of both instruments; and also by certified methods using *Vibrio fisheri* (Microtox®) and *Daphnia magna* (planktonic crustaceans).

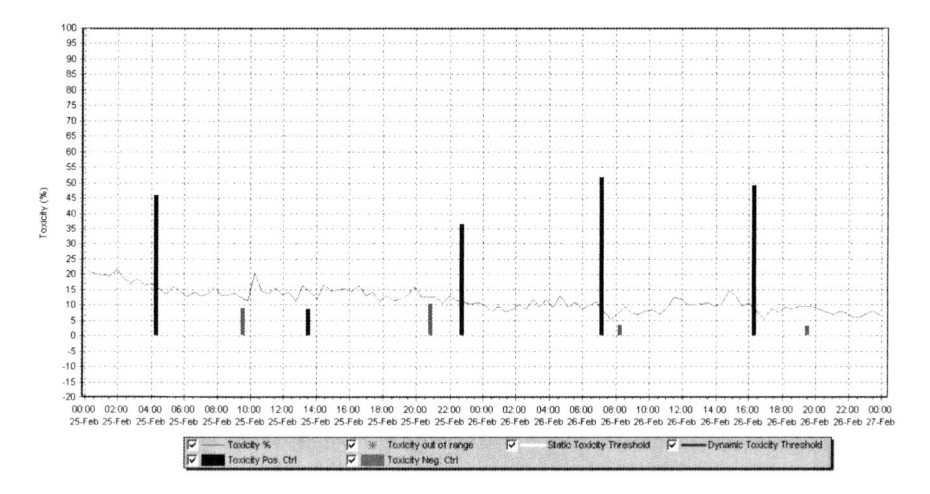

Figure 3 *Toxicity background levels corresponding to episodes of high turbidity*

Figure 4 shows that, although some differences can be seen comparing the two versions of equipment, no significant inhibition was shown so no toxicity of any of the samples could be reported. Checking dispersion of data, repeatability was better at the biomonitor in validation, as it is a more mature version that the new prototype. Analysis performed with Microtox® and *Dahpnia magna* showed as well no toxicity.

The equipment shows the advantage of detecting global toxicity of some of the priority pollutants (metals and pesticides) that causes inhibition in *Vibrio fisheri* bacteria. It can operate as an alert system for intake protection and the maintenance effort is low (once a week). Main disadvantage of the equipment is that bacteria are not sensitive enough to detect presence of toxic compounds at the levels needed to comply with legislation. To contribute to solve this problem, a Solid Phase Extraction (SPE) concentrator prototype is also being tested to increase sensitivity of this biomonitor to toxic substances.

3.4 Future work

Once the equipments have been validated for real sample analysis, experiments will be performed for a selection of ten target compounds. These substances have been selected according to their occurrence in Llobregat River and the potential to show toxicity. The following compounds will be used for testing: Terbutylazine, Diazinon, Dimethoate, Diuron, Propanil, MCPA, Nonylphenol, LAS, Triclosan, Diclofenac.

3.4.1 Recovery tests for SPE concentrator

In order to test performance of the SPE module for increasing concentration of our target compounds so they can show toxicity when analysed by biomonitor, the following experiment has been designed. HPLC grade water and surface water (filtered 0.45μm) will be spiked with a standard mixture of the ten target compounds. Samples will be spiked at 0.2 mg L^{-1} and 0.2 ng L^{-1} so after concentration (from 500 mL to a final volume of 10 mL), final concentrations are expected to be 10 mg L^{-1} and 10 ng L^{-1}. Two different cartridges will be tested: original ones and self-prepared (using OASIS HLB Waters® material). Moreover, three replicates plus blank will be prepared for each situation.

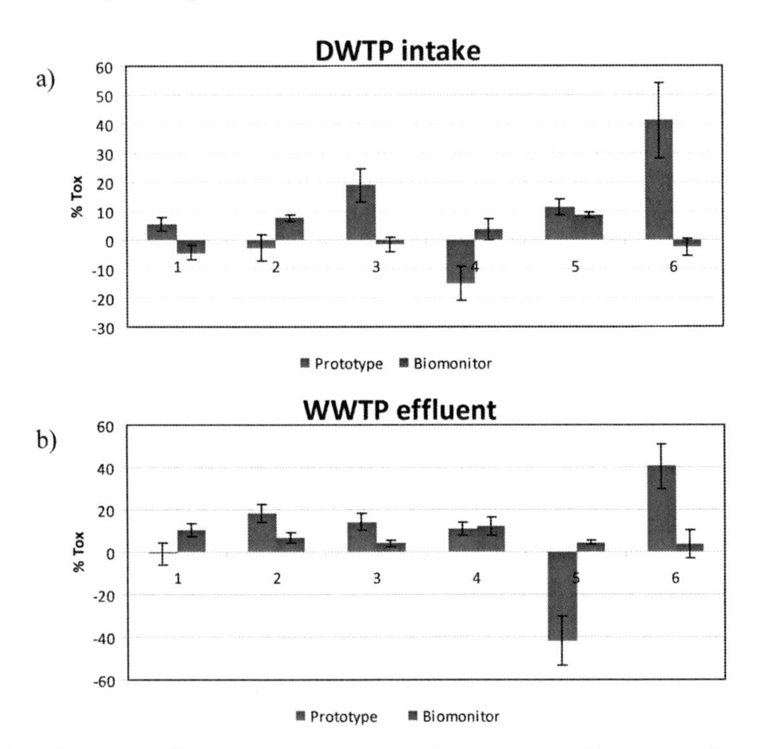

Figure 4 *Comparison between two equipments (prototype and biomonitor) for surface waters (a) and treated wastewaters (b). Samples are numbered from 1 to 6 corresponding to consecutive weeks*

3.4.2 Dose response curves

Concentration that causes 50% of inhibition (EC50) will be calculated according to "dilution series procedure" (Toxcontrol) for each target compound. In this procedure, different concentrations of the same compound are tested automatically to obtain the curve concentration-response that will show information for the EC50 calculation. Due to the low solubility of some of target compounds, solutions will be prepared in DMSO 0.2%. Such a low concentration of DMSO has been checked to show no response for *Vibrio fisheri*.

3.4.3 Toxicity tests for the target compounds

Taking into account information on recovery and toxicity obtained from previous tests, experiments will be designed for testing response of toxicity monitor when facing different mixtures of our target compounds after pre-concentration. Synergistic or antagonists effects could be checked. The influence of matrix in the response will be also an important issue to assess.

4 CONCLUSIONS

Water resources quality monitoring is a mandatory issue for ecological and health reasons. New tools should be integrated in water quality monitoring programmes. VIECO project develops methodology and validate technology previous to their routine use. Special characteristics of case study (Llobregat River) pose a challenge to the implementation of some technologies, specially the ones operating on line.

Weather condition can alter parameters like turbidity making difficult the measurement of this or other parameters when previous calibration is needed or when optical measures are implied (absorbance or luminescence).

Sensitivity is another important obstacle to overcome when on line instruments are used. Pre-treatment of sample can augment concentration of pollutants but could increase response time or imply implementation of some manual processes.

References

1 Directive 2000/60/EC of the European Parliament and of the Council of 23 October 2000 establishing a framework for Community action in the field of water policy. Official Journal of the European Communities 22.12.2000 L 327/1-72
2 Directive 2008/105/EC of the European Parliament and of the Council of 16 december 2008 on environmental quality standards in the field of water policy. Official Journal of the European Union 24.12.2008 L 348/84-97
3 Council Directive 98/83/EC of 3 November 1998 on the quality of water intended for human consumption. Official Journal of the European Communities 5.12.98 L 330/32-54
4 World Health Organization (WHO) Guidelines for Drinking-water Quality. Third Edition. Volume 1. Recommendations. 2004
5 ISO 22000:2005, Food safety management systems – Requirements for any organization in the food chain
6 M. Kuster, M.J. López de Alda, M.D. Hernando, M. Petrovic, J. Martín-Alonso, D. Barceló, *Journal of Hydrology*, 2008, **358**, 112
7 R. López-Roldán, M. López de Alda, M. Gros, M. Petrovic, J. Martín-Alonso, D. Barceló, *Chemosphere*, 2010, **80** (11), 1337
8 R. Céspedes, S. Lacorte, D. Raldúa, A. Ginebreda, D. Barceló, B. Piña, *Chemosphere*, 2005, **61** (11), 1710

Acknowledgement

This work is being financed by Spanish Ministry of Environment through VIECO project (009/RN08/01.1)

RAPID CONFIRMATION OF MICROBIOLOGICAL ALERTS USING OFF-LINE MOLECULAR METHODS

C.W. Keevil and S.A. Wilks

School of Biological Sciences, University of Southampton, Highfield Campus, Southampton SO17 1BJ, UK.

1 INTRODUCTION

In the event of a microbiological alert, either due to the triggering of an on-line sensor or malicious threats made to the news media, there is an urgent need to rapidly confirm the identity and quantity of the pathogen or indicator species involved in the emergency. If positive, then immediate decisions can be taken with confidence to quarantine the water and/or issue a boil notice to protect public health. A major problem is that there are several confounding factors which influence the ability to detect the contaminant. For example, during the time taken to respond to an alert the microbe may be severely diluted and requires a concentration step(s) such as large scale filtration of water. Alternatively, the microbe may adsorb to the walls of the pipe and associated corrosion deposits, biofilm or sediment.[1] Moreover, the microbe may be stressed from exposure to the low nutrient water or chlorine treatment. Existing isolation and culturing methods are time consuming, taking as long as 2-3 months for slow growing pathogenic species, or may not work for those environmentally stressed species in a viable but non-cultivable (VBNC) state.[2]

Consequently, to overcome such problems, rapid identification will most likely entail application of off-line detection methodology, utilising state of the art molecular techniques, for contaminants in water and on distribution network pipe surfaces following the probable adsorption of some of the agent. The methods should be appropriate for indicator microorganisms (following an accidental contamination) or biothreat agents (following an act of terrorism). Clearly, methods should be suitable for stressed or VNBC cells, and be not only sensitive (if necessary, requiring concentration steps but avoiding amplification artefacts) but also rapid since government agencies require a fast, reliable confirmation of the threat. Therefore, methods should be specific and sensitive to avoid false-positive or false-negative identification, respectively, in suspect situations which could confuse emergency responders and panic the local population. An important requirement is that such methods are robust: it is not sufficient that such methods work under laboratory conditions, they must be capable of working in diverse water chemistries such as high humic acid or iron concentrations, and with complex pipe deposits. If necessary, the agent must be successfully separated from the pipe deposits before analysis can be completed, requiring improved methodologies for sample preparation.

This chapter will review progress of rapid detection and viability assessment methods, including a critical assessment of the importance of the VBNC environmentally stressed phenotype in normally treated distribution systems and following remediation strategies. This knowledge will provide reassurance to the authorities that pathogens have indeed been removed and/or killed following decontamination interventions, and that the water distribution system can be signed off as being fit to return to normal use.

2 DETECTION OF INDICATORS OF MICROBIAL INGRESS

Indicators of faecal contamination have been well established in legislation for routine analysis of potable water, including faecal contaminants such as *Escherichia coli*, enterococci and *Cryptosporidium parvum* (mostly as a parameter of the water purification process), or other contaminants such as coliforms. The methods include classical culture techniques and enzyme detection, such as the Colilert and Colisure systems.[2] More recently, rapid antibody and nucleic acid based methods for detection are becoming available.

There are five approved or pending methods for analysis of *E. coli* in drinking water. These include i) Lactose TTC Tergitol Method (ISO 9308-1); ii) Colilert®-18/Quanty-Tray (IDEXX, UK National Standard Method W 18); iii) Membrane Lauryl Sulphate Agar (MLSA) (NEN 6553); iv) Membrane Lauryl Sulphate Broth (MLSB) (UK National Standard Method W 2) and v) Chromocult® Coliform Agar (Merck). Of note, none of the cultivation methods provides results faster than 1-2 days from the start of the analysis.

Rapid methods for *E. coli* analysis include flow cytometry which has been used to analyze *E. coli* in source water [3,4] and also in drinking water.[5] This technique can be combined with viability assays to assess the efficacy of the disinfectant process.[6] The technique can be considered as promising for monitoring the water distribution system. However, a problem with analysis of biofilm might be the undissolved autofluorescent particles that can be present. Under the microscope they are discernible from the cells but the cytometer might detect them as cells and thus increase the cell number. An alternative technique is solid phase cytometry which captures microbes on a membrane and then scans for fluorescently labelled cells e.g. Chemscan and provides rapid presumptive quantification.[2] This can be combined with a confirmatory step by recording the coordinates of each positive event and transferring the membrane to a fluorescence microscope which can then confirm the cell morphology of the putative positive signal at each designated coordinate on the filter.

The real time or quantitative PCR (qPCR) technique has been widely applied for drinking water and wastewater analyses of *E. coli*.[7-9] Following a comparative methods study, in which cell suspensions were quantified by epifluorescence microscopy, flow cytometry, qPCR and culture methods, it was concluded that the latter considerably underestimated the cell count whereas the former were comparable with each other and all could be recommended for practical use.[10] While good progress has been made with developing broth pre-enrichment and agar resuscitation methods for stressed cells, and even amoebal resuscitation of certain VBNC pathogens,[11] these require at least 24 hours if not several days to obtain a result. For rapid identification and response to a microbial contaminant it is clear that molecular methods should be used.

3 DETECTION USING MOLECULAR METHODS

3.1 PCR/qPCR

The PCR reaction has 3 stages, comprising i) denaturing of DNA from double to single-stranded; ii) annealing of primers to complementary sites on target DNA and iii) extension of DNA away from primers by addition of nucleotides resulting in production of double-stranded DNA. By cycling this process, it possible to amplify the signal over a million times (Figure 1).

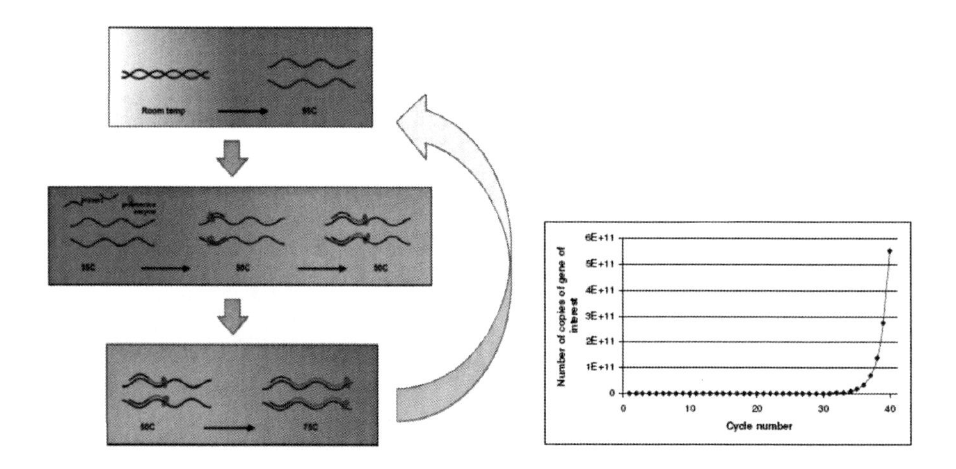

Figure 1 *The PCR cycle and amplification of DNA*

PCR assays exhibit very high level of specificity BUT have limitations. Firstly, although the PCR reaction can amplify a single copy of DNA by a million-fold in 2 hours, it may still need culture enrichment prior to this. This problem is generally caused by interference from other materials in the sample. Secondly, the assay is not quantitative: amplification usually saturates during the later cycles and the amount of a specific sequence in the final amplification product is not always proportional to the amount of that sequence in the original sample. Thirdly, PCR cannot provide a measure of viability, so samples have to be pre-enriched and then the grown cells are measured. This clearly delays the time to obtain positive results and also presents a problem if the cells are VNBC and cannot be pre-enriched.

The development of real-time PCR allowed a signal to be quantified (qPCR) which is a great advantage over the standard PCR technique. qPCR quantifies the target sequence by comparing the amount of fluorescence marker bound to the increasing copies of DNA (Figure 2). The horizontal line on the graph represents a “threshold” set by the user. The point at which the amplification plot crosses this threshold is known as the Ct (cross threshold) value. The lower the Ct value for a sample the greater the starting amount of DNA in the sample. Thus deduce which sample contained the greatest amount of the DNA of interest by the Ct value. Blue sample Ct value=23; green sample Ct value=28. Therefore the blue sample contained 32 (2^5) times more of the gene of interest than the green sample.

Use of a single reaction tube reduces contamination problems and therefore reduces false positive or negative results. Despite the clear advantages of qPCR over PCR, while the use of qPCR can overcome the problem of quantification it is still not possible to measure cell viability. In this sense, it can only be used as a presence/absence test.

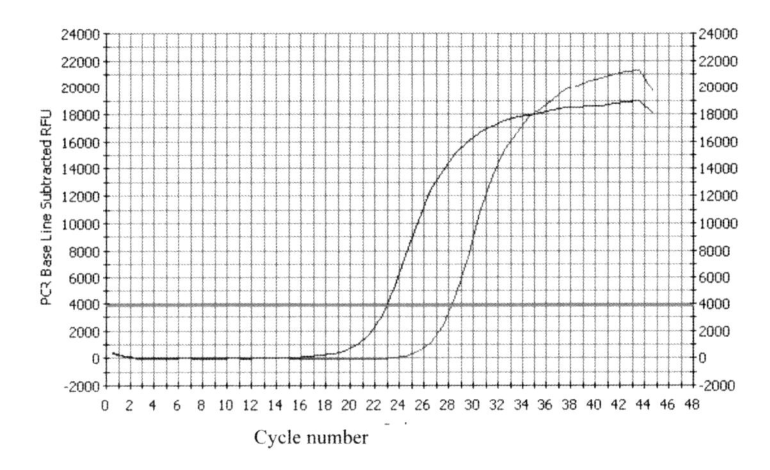

Figure 2 *The qPCR amplification plot showing the number of PCR cycles (1 cycle = 90°C, 50°C, 72°C) against the increasing fluorescence.*

An important improvement in qPCR has been the development of monoazide inhibition of the PCR in dead cells. This technique initially relies on qPCR to quantify microbes, providing a measure of population density. However, for viability assessment it incorporates a monoazide: initial studies used ethidium monoazide (EMA) a large molecule which can only enter cells with a damaged membrane. It will then form a nitrene when it undergoes photolysis with visible light and irreversibly binds to DNA, causing a conformational change which inhibits the PCR/qPCR reaction. However, it became apparent that EMA is too small, giving a false impression of the % dead cells. Therefore, propidium monoazide (PMA) was used instead (Figure 3).[12]

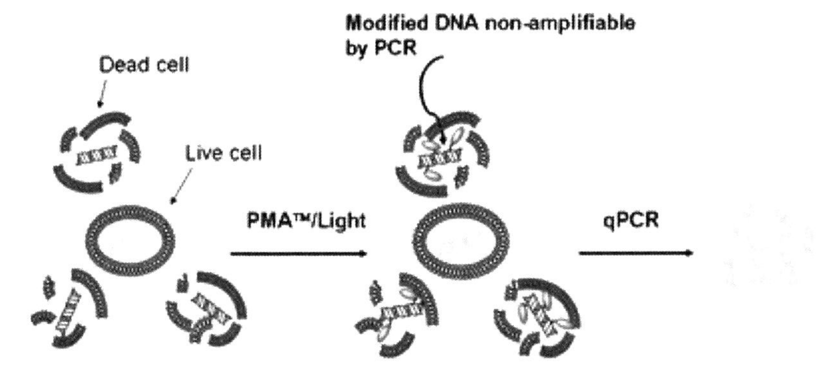

Figure 3 *The PMA-qPCR reaction.*

We are now investigating this technique for *E. coli* O157 and other agents on the biothreat list etc by designing and evaluating specific sets of primers and probes. An early example is shown in Figure 4 where we have compared the detection of populations of

100% live cells, 100% dead cells and a mixture of 10% live/90% dead. Good discrimination can be seen between the different populations. Indeed, it is feasible to discriminate less than 1% viable cells at concentrations as low as 100 cells.

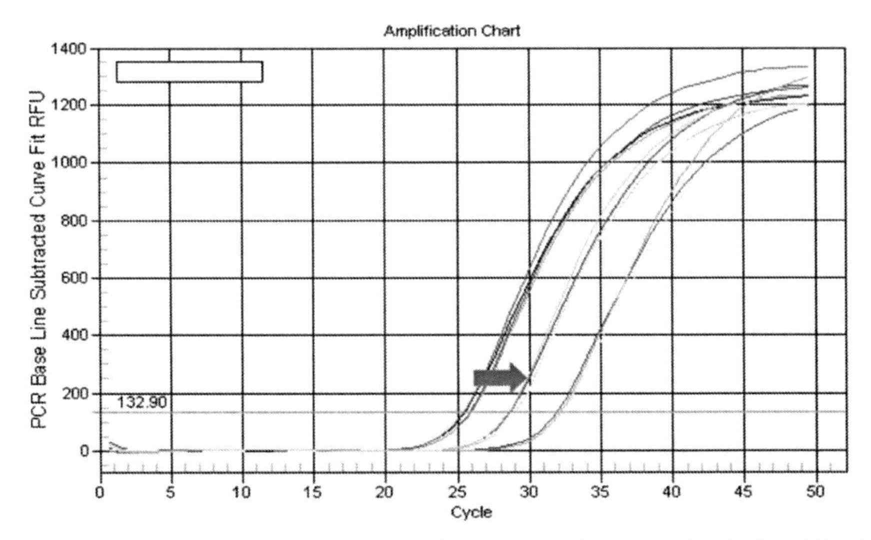

Figure 4 *The PMA-qPCR reaction (arrow shows the middle plot of 10% live/90% dead cells while the adjacent left and right plots show 100% live and 100% dead cells, respectively.*

3.2 Fluorescence *in situ* hybridisation

Fluorescence *in situ* hybridisation (FISH) assays have been widely used e.g. for specific 16S rRNA labelling of bacteria but they have a number of limitations which have restricted their routine application. Generally, the technique relies on the use of DNA probes which target a specific region on the DNA/RNA of the organism of interest. The main problem with DNA probes is that they have a strong requirement for highly specific hybridisation conditions making it difficult to use them on complex samples from the environment where there could be changes in pH, redox, etc. Moreover, multiplex assays are actually a compromise because each target in a different microbial species has its own specific hybridisation conditions, so the likelihood is only one target will be optimally detected in a defined set of hybridisation conditions while other targets will not be at their optimum. DNA probes are also prone to charge effects due to the strong negative charge of the phosphate backbone, making them less easy to enter cells and also bind strongly to corrosion deposits.

By contrast, peptide nucleic acid (PNA) probes for FISH have many advantages over DNA probes.[13-16] PNA are synthetic DNA analogues with a 2-aminoethyl-glycine linkage replacing the phosphodiester backbone found in DNA (Figure 5).

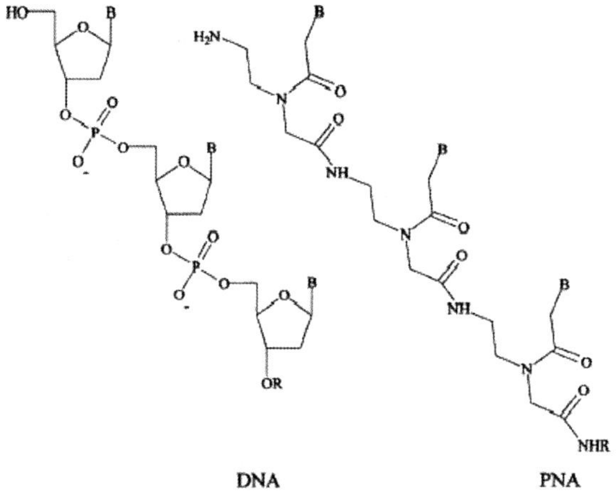

Figure 5 *Structure of charged DNA and neutral PNA probes*

PNA probes are normally designed to target areas on the 16S rRNA molecule. This molecule is chosen as a target due to its high copy number, obviating the need for external amplification processes such as PCR. PNA probes are shorter than standard DNA probes and so are able to access often inaccessible areas of this molecule (Figure 6);[17] increasing their specificity.

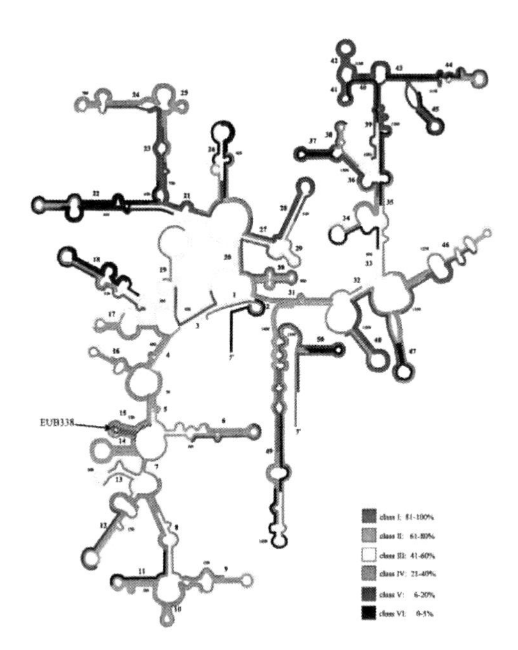

Figure 6 *Structure of E. coli 16S rRNA showing high and low probe binding sites*

Other attributes of PNA probes include i) capable of sequence-specific recognition of DNA and RNA, obeying Watson-Crick hydrogen bonding rules; ii) exhibit greater thermal stability; iii) not susceptible to hydrolytic cleavage and iv) require less stringent hybridisation conditions. These are useful properties when examining complex environmental samples. In particular, we have been able to use PNA-FISH assays for the detection of specific pathogens. Examples include VBNC *Campylobacter coli* (Figure 7a)[14] and *Mycobacterium avium* in drinking water (Figure 7b)[15] and VBNC *Helicobacter pylori* in a drinking water biofilm (Figure 8).[18]

Figure 7 *PNA-FISH detection of C. coli (A) and M. avium (B) in drinking water*

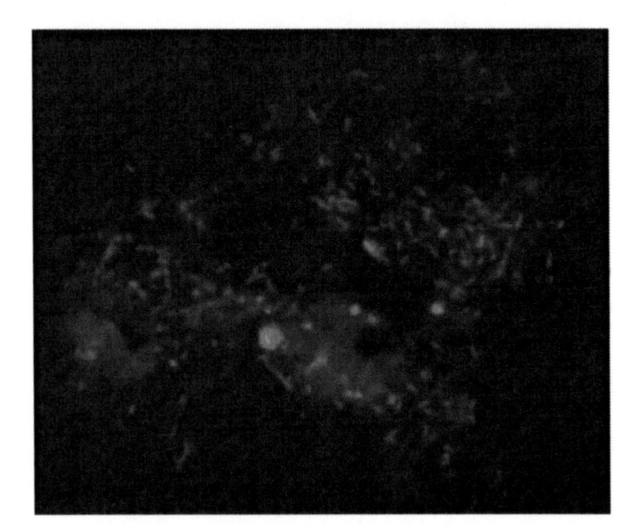

Figure 8 *PNA-FISH detection of H. pylori in a polymicrobial drinking water biofilm*

Despite the many advantages of PNA-FISH the technique, like PCR, cannot provide accurate information on the viability of detected target pathogens. However, an old technique called the cell elongation assay or direct viable count (DVC), originally described by Kogure *et al.*,[19] has been modified for use in drinking water and biofilms and incorporated into the PNA-FISH assay. This requires the application of an antibiotic, pipemidic acid, which allows cell growth in samples incubated with a low concentration of nutrient such as R2 medium, to avoid nutrient shock, but inhibits cell division, resulting in highly extended bacteria which can be easily visualised using fluorescent DNA binding dyes such as DAPI. Specificity can be obtained using FISH to provide a combined methodology of specificity and viability. This approach was used in a study of *E. coli* in European drinking water supplies where viable *E. coli* could be detected in biofilm samples recovered from distribution network while these could not be recovered by conventional culture, Colilert or Colisure techniques (Figure 9).[20] The DVC technique is relatively rapid but probably more appropriate for confirmation of remediation strategies following accidental or deliberate ingress of a pathogen, rather than routine monitoring.

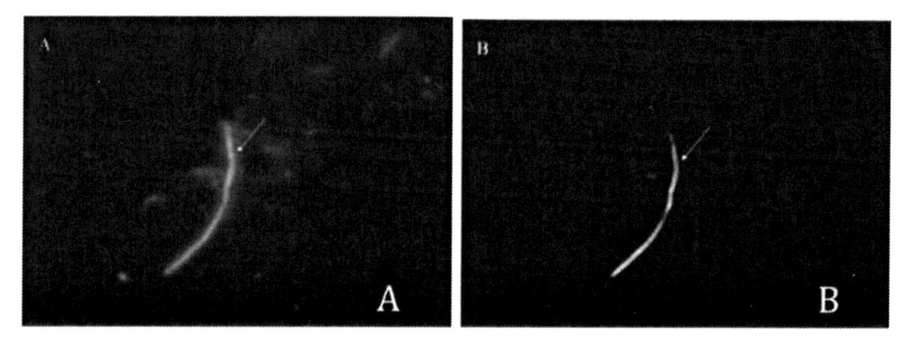

Figure 9 *DVC-PNA-FISH detection of viable E. coli in a drinking water biofilm using DAPI to show total viable cell elongation (A) or a 16S rRNA PNA probe (B) for specific detection.*

A similar approach can be undertaken to identify viable spores of *Bacillus* spp. For example, we have recovered drinking water biofilm containing *Bacillus* spores and incubated them in a low nutrient medium to promote vegetative growth. These can be seen in resuspended biofilm samples captured on filter media using a specific PNA probe targeting a region of the 16S rRNA (Figure 10a). If the assay is repeated in the presence of pipemidic assay, then extended viable cells can clearly be seen in this modified DVC assay (Figure 10b), showing that the assay is applicable to filtered water samples and not just biofilm or deposit samples. Similarly, viable *Bacillus* spores can be seen directly in drinking water biofilm samples without the need for resuspension (Figure 11). Indeed, it is apparent that the extent of cell elongation is greater in the biofilm during the DVC assay.

We are now applying these techniques to studies of remediating drinking water and associated biofilms with high concentrations of chlorine, monochloramine and chlorine dioxide. For example, the initial data indicate that *E. coli* O157 can become non-cultivable in the presence of very high concentrations of disinfectant but still appears viable according to the DVC assay. These results should have a profound impact on our assessment and understanding of disinfection and pathogen clearance in a drinking water network.

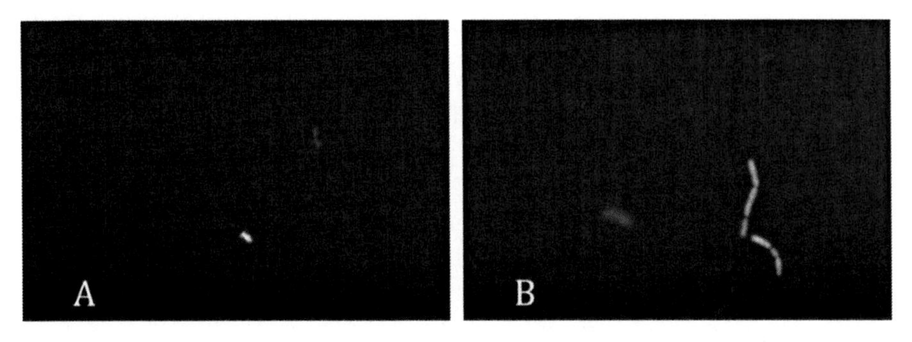

Figure 10 *PNA-FISH detection of viable Bacillus spores in a resuspended drinking water biofilm using the direct PNA-FISH assay (A) and the DVC modification of the PNA-FISH assay (B).*

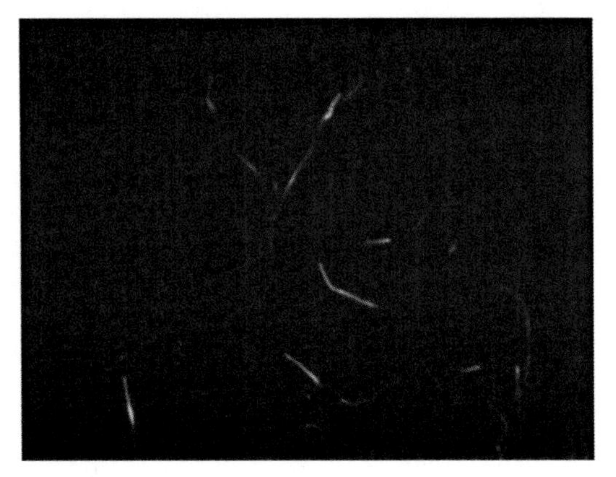

Figure 11 *DVC-PNA-FISH detection of viable Bacillus spores directly in a drinking water biofilm.*

4 CONCLUSIONS

Good progress is now being made for rapid molecular detection and viability assessment methods, including the importance of the VBNC environmentally stressed phenotype. These methods are robust, sensitive and specific, satisfying the demanding criteria outlined at the start of this chapter to avoid false positive or false negative information and capable of providing quantitative data. As such, they should be suitable for normally treated distribution systems when accidental or deliberate microbial ingress occurs and following remediation strategies. This knowledge will provide reassurance to the authorities and the general public that pathogens have indeed been removed and/or killed following decontamination interventions. As such, the water distribution system can then be signed off as being fit to return to normal use.

5 ACKNOWLEDGEMENT

This work has been undertaken as part of a research project entitled SECUREAU which is supported by the European Union within the Seventh Framework Programme under Theme

10 Security, Project n°217976. There hereby follows a disclaimer stating that the authors are solely responsible for the work, it does not represent the opinion of the Community and the Community is not responsible for any use that might be made of data appearing herein.

References

1. C.W. Keevil. Pathogens and metabolites associated with biofilms. In *Biofilms in the Aquatic Environment,* ed. C.W. Keevil, A. Godfree, D. Holt and C.S. Dow, Cambridge: Royal Society of Chemistry, 1999, 145-152.
2. C.W. Keevil. Suitability of microbial assays for potable water and wastewater applied to land. In *Rapid Detection Assays for Food and Water,* ed. S.A. Clark, C.W. Keevil, K.J. Thompson and M. Smith, Cambridge: Royal Society of Chemistry, 2001, 3-26.
3. C. Sakamoto, N. Yamaguchi and M. Nasu. Rapid and simple quantification of bacterial cells by using a microfluidic device. *Appl. Environ. Microbiol.*, 2005, **71**, 1117-21.
4. Y. Tanaka, N. Yamaguchi and M. Nasu. Viability of *Escherichia coli* O157:H7 in natural river water determined by the use of flow cytometry. *J. Appl. Microbiol.*, 2000, **88**, 228-36.
5. C. Sakamoto, N. Yamaguchi, M. Yamada, H. Nagase, M. Seki and M. Nasu. Rapid quantification of bacterial cells in potable water using a simplified microfluidic device. *J. Microbiol. Methods*, 2007, **68**, 643-7.
6. M. Berney, H.U. Weilenmann and T. Egli. Flow-cytometric study of vital cellular functions in *Escherichia coli* during solar disinfection (SODIS). *Microbiology*, 2006, **152**, 1719-29.
7. E. Frahm and U. Obst. Application of the fluorogenic probe technique (TaqMan PCR) to the detection of *Enterococcus* spp. and *Escherichia coli* in water samples. *J. Microbiol. Methods*, 2003, **52**, 123-31.
8. A.M. Ibekwe, P.M. Watt, C.M. Grieve, V.K. Sharma and S.R. Lyons. Multiplex fluorogenic real-time PCR for detection and quantification of *Escherichia coli* O157:H7 in dairy wastewater wetlands. *Appl. Environ. Microbiol.*, 2002, **68**, 4853-62.
9. S. Tantawiwat, U. Tansuphasiri, W. Wongwit, V. Wongchotigul and D. Kitayaporn. Development of multiplex PCR for the detection of total coliform bacteria for *Escherichia coli* and *Clostridium perfringens* in drinking water. *Southeast Asian J. Trop. Med. Public Health*, 2005, **346**, 162-9.
10. P.S. Chen and C.S. Li. Real-time quantitative PCR with gene probe, fluorochrome and flow cytometry for microorganism analysis. *J. Environ. Monit.*, 2005, **7**, 257-62.
11. S. Giao, S.A. Wilks, N.F. Azevedo, M.J. Vieira, and C.W. Keevil. Validation of SYTO 9/Propidium Iodide uptake for rapid detection of viable but non-cultivable *Legionella pneumophila. Microbial Ecol.*, 2009, **58**, 56-62.
12. A. Nocker, C.Y. Cheung and A.K. Camper. Comparison of propidium monoazide with ethidium monoazide for differentiation of live vs. dead bacteria by selective removal of DNA from dead cells. *J. Microbiol. Methods*, 2006, **67**, 310-20.
13. N.F. Azevedo, M.J. Vieira and C.W. Keevil. Establishment of a continuous model system to study *Helicobacter pylori* survival in potable water biofilms. *Water Sci. Technol.*, 2003, **47**, 155-160.
14. M.J. Lehtola, C.J. Loades and C.W. Keevil. Advantages of peptide nucleic acid oligonucleotides for sensitive site directed 16S rRNA *in situ* hybridisation (FISH) detection of *Campylobacter jejuni*, *Campylobacter coli* and *Campylobacter lari. J. Microbiol. Methods*, 2005, **62**, 211-219.

15. M.J. Lehtola, E. Torvinen, I.T. Miettinen and C.W. Keevil. Fluorescence *in situ* hybridization (FISH) using peptide nucleic acid probes for rapid detection of *Mycobacterium avium* subsp. *avium* and *Mycobacterium avium* subsp. *paratuberculosis* in potable water biofilms. *Appl. Environ. Microbiol.*, 2006, **72**, 848-853.
16. S.A. Wilks and C.W. Keevil. Targeting species-specific low affinity 16S rRNA binding sites using peptide nucleic acids for the detection of *legionellae* in biofilms. *Appl. Environ. Microbiol.*, 2006, **72**, 5453-5462.
17. B.M. Fuchs, G. Wallner, W. Beisker, I. Schwippl, W. Ludwig and R. Amann. Flow cytometric analysis of the in situ accessibility of *Escherichia coli* 16S rRNA for fluorescently labeled oligonucleotide probes. *Appl. Environ. Microbiol.*, 1998, **64**, 4973-82.
18. S. Giao, S.A. Wilks, N.F. Azevedo, M.J. Vieira, and C.W. Keevil. Effect of chlorine on the incorporation of *Helicobacter pylori* into drinking water biofilms. *Appl. Environ. Microbiol.*, 2010, **76**, 1669–1673.
19. K. Kogure, U. Simidu and N. Taga. A tentative direct microscopic method for counting living marine bacteria. *Can. J. Microbiol.*, 1979, **25**, 415-20.
20. T. Juhna, D. Birzniece, S. Larsson, D. Zulenkovs, A. Shapiro, N.F. Azevedo, F. Menard-Szczebara, S. Castagnet, C. Feliers and C.W. Keevil. Detection of *Escherichia coli* in biofilms from pipe samples and coupons in drinking water distribution networks. *Appl. Environ. Microbiol.*, 2007, **73**, 7456-64.

IMPROVING QUALITY AND SAVING DOLLARS USING REAL-TIME ONLINE WATER QUALITY MONITORING

K. Thompson, G. Jacobson and K. Chamberlain

CH2M HILL World Headquarters, 9191 South Jamaica Street, Englewood, CO 80112, USA

1 INTRODUCTION

Online water quality monitoring (OWQM) stations provide utility operators and decision-makers with continuous, real-time water quality data from key locations in their system, such as source water, raw water conveyance, treatment plants, and the distribution system. Implementation of OWQM stations is a state-of-the-art means for maintaining water quality, warning of potential contamination, and helping to prevent public health problems, damage to the system, compromised regulatory compliance, lack of water availability, and loss of public confidence. Several utilities across the United States have implemented advanced OWQM stations in their distribution systems, optimizing treatment operations and gaining the benefits listed above, as well as the associated cost savings.

2 DESIGN CRITERIA

For early warning of accidental or intentional contamination, the U.S. Environmental Protection Agency has identified twelve classes of contaminants of concern (shown in Figure 1). These include petroleum products, pesticides, inorganic compounds, toxins, pathogens, and chemical warfare agents. Continuous monitoring of chlorine residual, total organic carbon, and conductivity has proven to be an effective way to detect most of these contaminant classes. At a minimum to protect distribution systems against intentional contamination events, OWQM stations should use sensors for these three parameters.

To provide the dual benefits associated with operational enhancements and regulatory compliance, several other parameters can be used to detect deterioration of distribution system water quality. For example, nitrification precursors may be identified by monitoring for total chlorine, nitrate, nitrite, free ammonia, and assimilable organic carbon. Utilities should consider which process parameters are most appropriate for monitoring in their systems to provide these dual benefits.

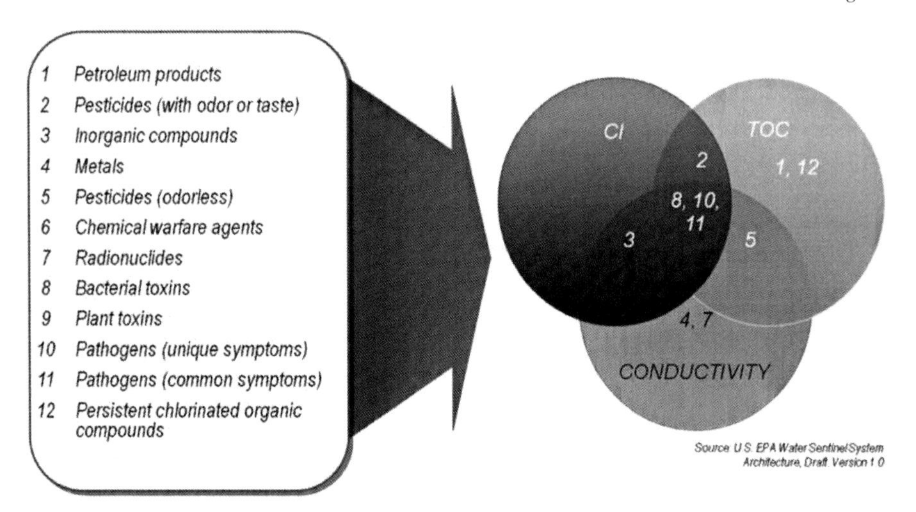

Figure 1 *US EPA Classes of Contaminants of Concern*

The ideal OWQM design attributes are:

- Rapid detection of system anomalies and provide notification for extended response time
- An alarm or report to set response activities in action
- Detection of a wide range of contaminants
- Indication of the contaminant source
- Affordability
- Robustness/reliability
- Low rate of false positive alarms and negative detections
- Remote operation
- Low skill level and limited training required to operate and maintain

When identifying the proper locations within the distribution system for OWQM station, criteria such as easy access, secure physical space, and readily available utilities (water, sewer, power, communication) should be considered. Potential station locations include:

- Service organizations (police stations, fire stations, schools)
- High-visible users (stadiums, arenas, shopping areas)
- Vulnerable populations (hospitals, daycare centers, nursing homes)
- Utility-owned facilities

3 EXAMPLES OF EMERGING AND STATE-OF-THE-ART OWQM ANALYZERS

During the last 10 years, there have been significant advancements in OWQM sensors that can provide sustainable solutions for the distribution system. Two examples of the new advancements in OWQM sensors that are becoming well-known in the marketplace are the

Intellisonde™ multi-parameter probe and the s::can spectro::lyser™ (Figures 2 and 3, respectively).

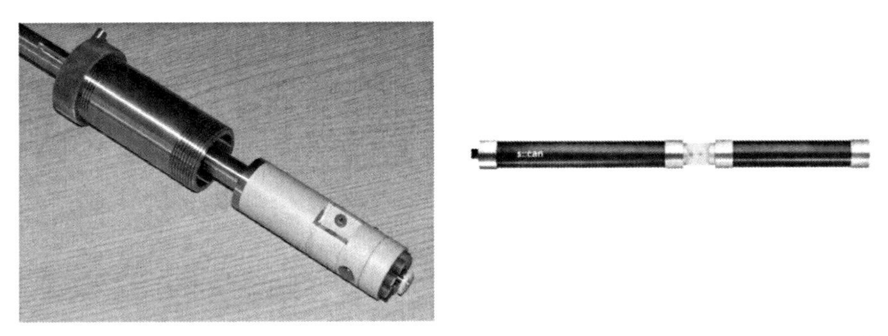

Figure 2 *Intellisonde™ Multi-parameter Probe* **Figure 3** *s::can spectro::lyser™*

The Intellisonde™ multi-parameter probe measures Cl_2 residual, conductivity, turbidity, pH, temperature, oxygen reduction potential, color, and flow, while the s::can spectro::lyser™ measures total organic carbon, ultraviolet spectral analysis, turbidity, dissolved organic carbon, NO_3, dissolved oxygen, NO_2, Cl_2, and organics. Table 1 provides a brief look at commonly used surrogate, non-TOC sensor suites available for use for OWQM in the distribution system. In addition, there are a number of single parameter sensors that are provided by a variety of vendors.

Table 1 *Non-TOC Sensor Suites*

Caption	Parameters Measured
Hach GuardianBlue Water Panel	pH, chlorine residual (total or free), conductivity, turbidity
ATI Q45WQ Water Quality Panel	pH, turbidity, chlorine residual (free or chloramines), conductivity, ORP
Intellitect Intellisonde™	pH, chlorine residual (free or chloramines), conductivity, turbidity, colour, ORP, flow, pressure
s::can Multi-Parameter Probe	pH, conductivity, chlorine residual (free or total), ammonia (free)

Figure 4 provides criteria useful in evaluating the non-total organic carbon (TOC) sensor suites listed in Table 1.

Vendor	Multi-parameter System Criteria								
	Footprint	Maintenance Cycle	Industry Experience	Capital Cost	U.S. Vendor Support	Reliability	Accuracy and Precision	Number of Parameters	Data Transfer Protocols
Hach	Poor or Unknown	Poor or Unknown	Excellent	$$	Excellent	Good	Excellent	4	Good
ATI	Poor or Unknown	Good	Excellent	$$	Excellent	Excellent	Excellent	4	Poor or Unknown
s::can multi-parameter	Excellent	Good	Excellent	$	Excellent	Excellent	Excellent	3-4	Excellent
Intellitect	Excellent	Excellent	Good	$	Poor or Unknown	Excellent	Good	5-9	Excellent

$ = <$10K

$$ = $10K-$25K

Poor or Unknown   Good   Excellent

Figure 4 *Evaluation Criteria for Non-TOC Sensor Suites*

Table 2 provides brief understanding of chemical and ultraviolet (UV) TOC analyzers.

Table 2 *Chemical and UV TOC Analyzers*

Manufacturer / Instrument	Parameters Measured
Hach Astro TOC (Chemical Based)	TOC
GE Sievers (Chemical Based)	TOC
OI Analytical (Chemical Based)	TOC
s::can spectro::lyzer	TOC, DOC, turbidity, nitrate, spectral fingerprint (250-700 nm), specific contaminants
Hach UVAS	UV-254
RealTech	UV-254

Figure 5 provides criteria useful in evaluating the chemical and UV TOC analyzers listed in the Table 2.

4 DUAL BENEFITS

Several utilities have begun to realize the dual benefits to be gained from the implementation of OWQM stations in their distribution systems. Not only do OWQM stations provide security with the early identification of possible contamination events, they are earning their keep by providing additional benefits that include operational optimization and regulatory compliance, which in turn helps utilities to save money.

Manufacturer	Capital Cost	Relative Complexity	Relative Maintenance	Relative Footprint
Hach Astro TOC	$$$	!!!	Poor	Poor
GE Sievers Series 900	$$	!!	Poor	Excellent
OI Analytical	$$	!!	Excellent	Excellent
s::can spectro::lyser	$$	!	Excellent	Excellent
Hach UVAS (*)	$	!	Excellent	Excellent
RealTech (*)	$	!	Excellent	Excellent

(*) UV-254 Only

$ = <$10K
$$ = $10K-$25K
$$$ = >$25

Poor / Good / Excellent

Figure 5 *Evaluation Criteria for Chemical and Ultraviolet TOC Analyzers*

4.1 Example 1

A military facility was automatically alerted when a change of treated water from the wholesale supplier was introduced into the system, creating a significant change in water quality. Early detection enabled the utility to quickly contact the wholesale supplier, identify the cause of the shift from its typical baseline water quality, and take appropriate treatment response actions. Figure 6 demonstrates the effect on turbidity of a flow reversal in the distribution system; Figure 7 demonstrates the effect on chlorine by introducing a new treated water supply. Both of these events would trigger an alarm to the operations staff, alerting them to an unexpected change in water quality in the distribution system.

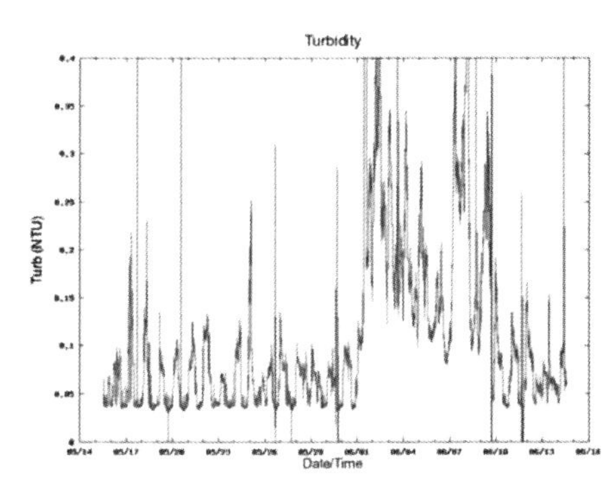

Figure 6 *High Turbidity Associated with Pipeline Scouring from Flow Reversal*

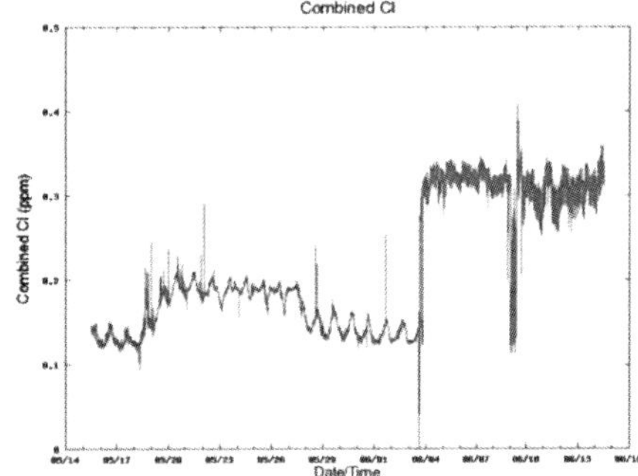

Figure 7 *Chlorine Change Associated with Treated Water Supply Change*

4.2 Example 2

A utility contractor installed and was operating a pilot ion exchange (IX) system at a reservoir pump station for treating nitrate. The IX system was designed to deliver treated IX water into the reservoir that mixed with conventionally treated water and discharge spent brine to the sewer. The IX pilot plant controller failed, discharging spent brine directly into the reservoir, which eventually traveled through the distribution system in a plug flow. The spent brine-contaminated water was detected by the OWQM station located at the storage reservoir and at another downstream OWQM station. The utility was able to identify the cause and extent of the water quality problem and implement corrective actions to prevent a recurrence. Figures 8 and 9 show the change in the spectral fingerprint from the s::can UV-Vis spectro::lyzer associated with the introduction of brine into the system.

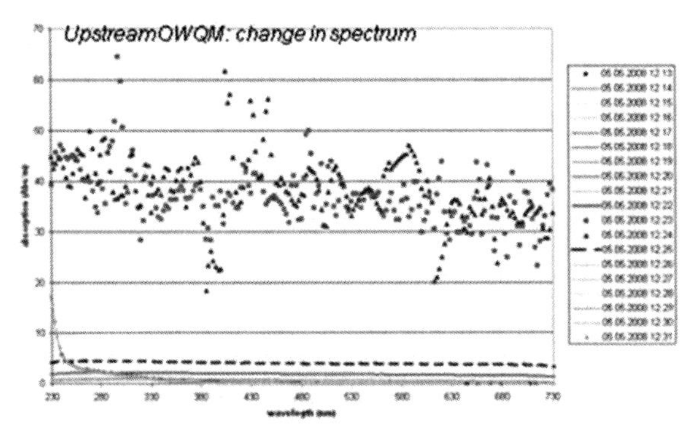

Figure 8 *Upstream Spectral Fingerprint Change*

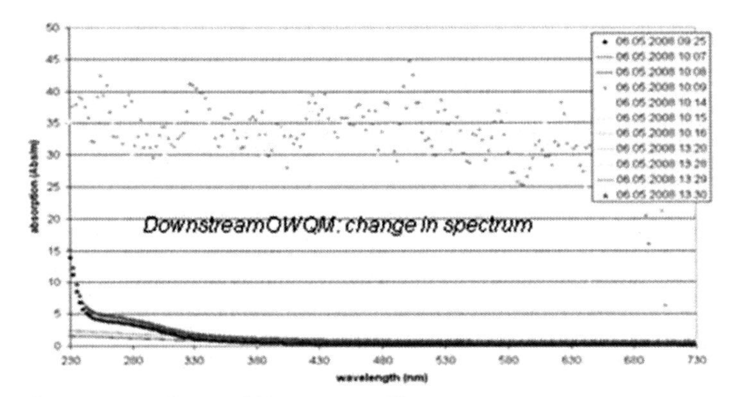

Figure 9 *Downstream Spectral Fingerprint Change*

4.3 Example 3

One utility experienced a failure with a caustic feed pump in the main water treatment plant, resulting in a finished water pH level markedly lower than normal. An OWQM alarm was activated as soon as the low pH alarm threshold was exceeded, and the utility was able to quickly identify the pump failure, correcting the problem within 2 hours, before water quality in the clearwell violated regulatory compliance limits. Figure 10 shows the pH profile before, during, and after the caustic pump failure.

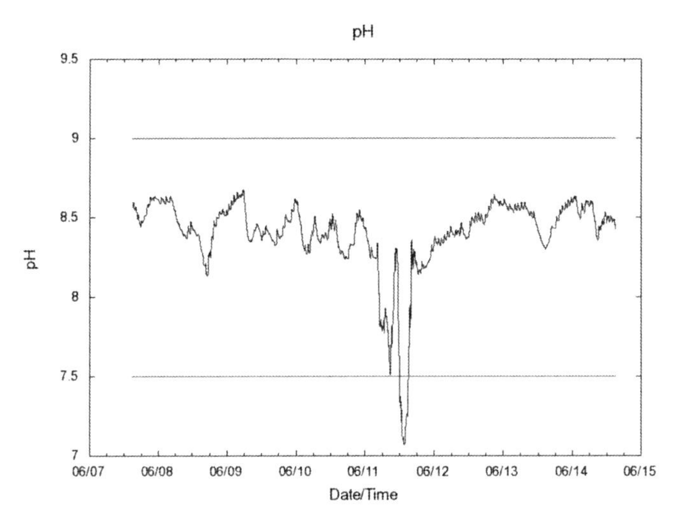

Figure 10 *Sodium Hydroxide Pump Failure*

4.4 Example 4

Using analysis of OWQM station historical water quality data (as shown in Figure 11), a utility was able to identify a distribution system problem attacking the ductile iron pipeline through the detection of low levels of iron oxide. The early identification and correction of the aggressive water quality problem saved this utility an estimated $20 million in early replacement costs of a relatively new section of distribution system piping.

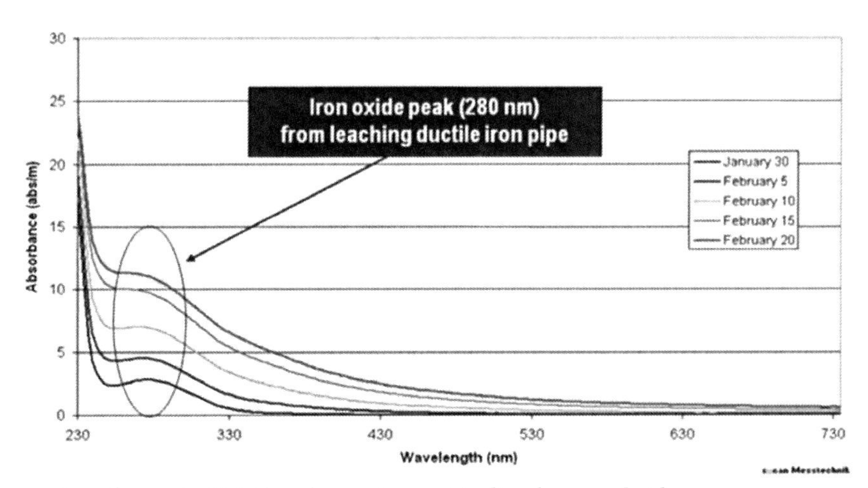

Figure 11 *Change in UV Absorbance Due to Fouling by Iron Oxide*

4.5 Example 5

Many utilities across the United States use chloramines as the primary disinfectant. These utilities struggle with the control of nitrification and the associated water quality problem such as loss of chlorine residual.

Figures 12 and 13 illustrate the water quality that an agency is receiving from its wholesale water company. The water has a reasonable chlorine residual (2.5-3 mg/L) and slight ammonia residual (0.5 mg/L).

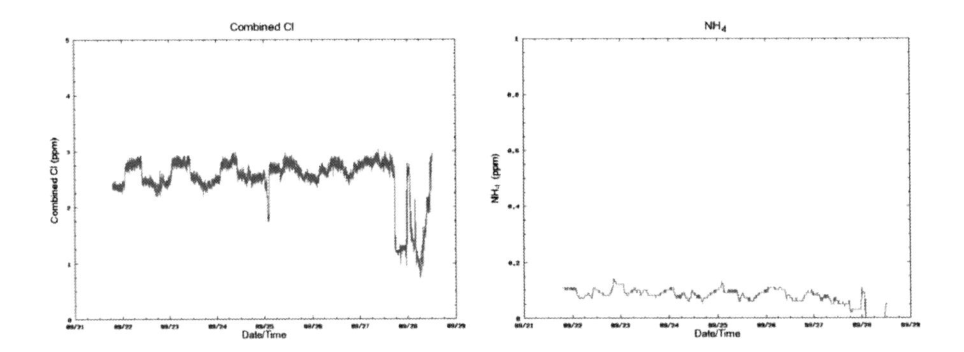

Figure 12 *Chlorine Residual Graph* **Figure 13** *Ammonia Graph*

Once the water has entered into the water agency's distribution system, the impacts of nitrification could be clearly seen with the loss of chlorine (Figure 14) and loss of ammonia (Figure 15). Once the flushing of the system was conducted to bring fresher water into the distribution system, both chlorine and ammonia levels started to increase. Based on field observations in a number of systems using chloramines as the primary disinfectant, loss of ammonia appears to be an excellent early warning indicator of nitrification events.

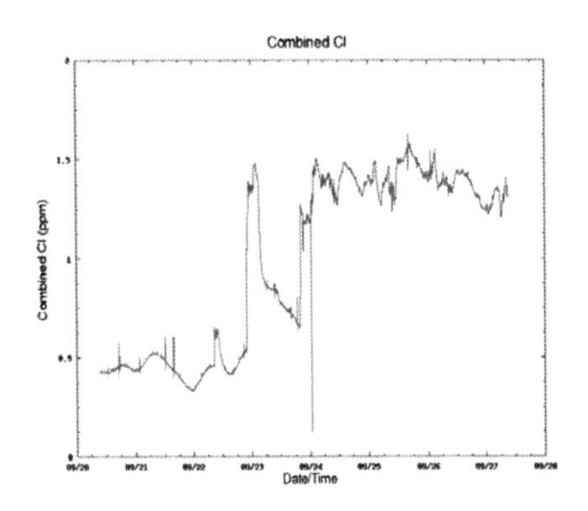

Figure 14 *Loss of Chlorine*

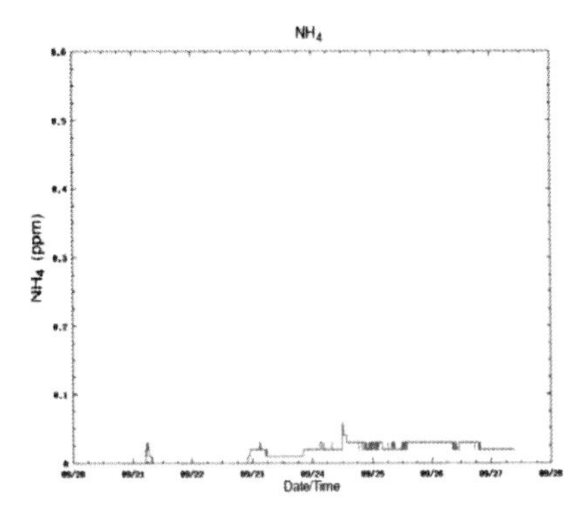

Figure 15 *Loss of Ammonia*

5 SECURITY EVENT MONITORING

In February 2008, the City of Glendale, Arizona hosted the National Football League Super Bowl (Championship Game). During the game and week prior to the event, over 250,000 people visited the venue, creating a highly viable target for intentional contamination (Figure 16).

Figure 16 *Photos of the 2008 Super Bowl Event Site*

In preparation for the Super Bowl, the City of Glendale developed an OWQM system to provide real-time monitoring and event detection prior to and during the event. Figure 17 illustrates two of the Web-based applications that were developed for the project. These applications allowed the City staff to view the OWQM data real time in multiple locations in its water distribution system from the Emergency Operations Center and externally by its consultant team for support.

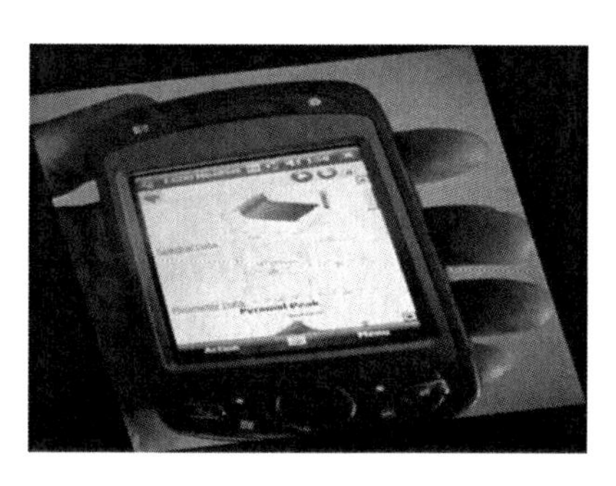

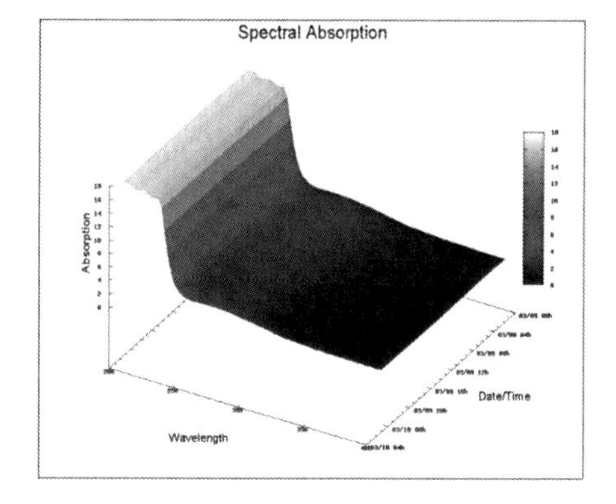

Figure 17 *Web-based Applications*

Game-time OWQM demonstrated that there was no contamination occurrence prior to and during the game (Figure 18). At the end of the game there was a change in the s::can spectro::lyzer spectral fingerprint. This was associated with a large surge of water associated with increased demand of water used for personal hygiene (Figure 19). This demonstrates the value of spectral fingerprint data for quickly identifying system scouring associated with sudden changes in increased water demand.

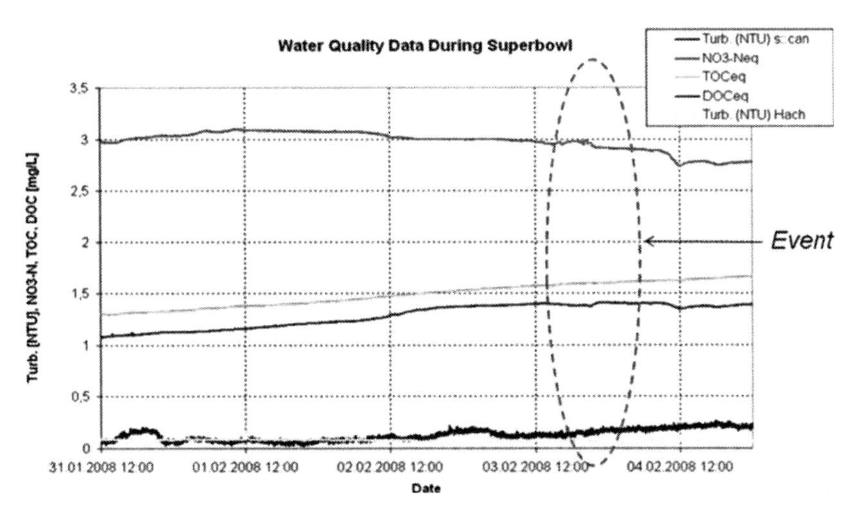

Figure 18 *Water Quality Data Collected During the 2008 Super Bowl*

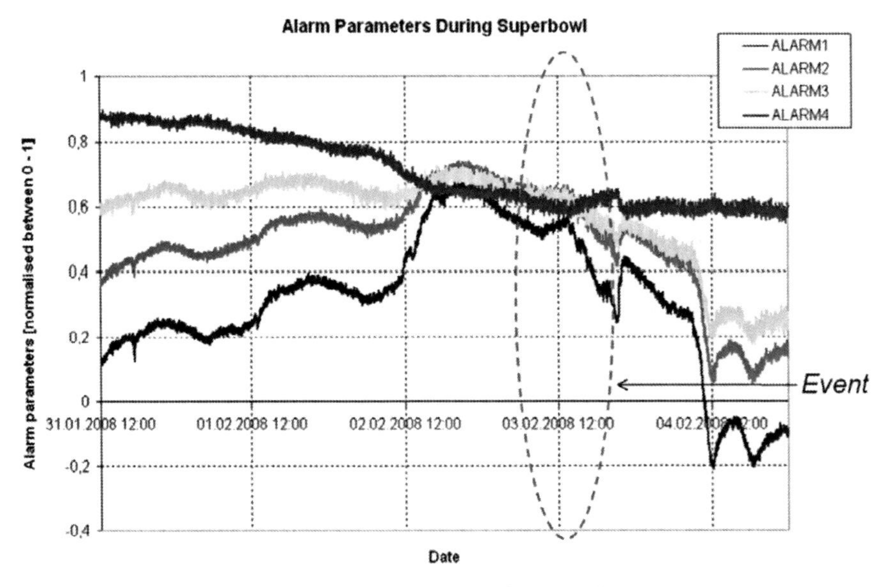

Figure 19 *Spectral Fingerprint Data – Example of Demand Change*

6 SUMMARY

Advanced OWQM programs for the distribution system provide multiple advantages for utilities:

- OWQM can provide environmental, public health, and economic benefits through identification of water quality changes in the distribution system.
- Several utilities have demonstrated operations and regulatory benefits, in addition to security benefits.
- Distribution system water contaminant monitoring systems are proven and sustainable.
- Monitoring for surrogate parameters is more often used than monitoring for specific contaminants, except for special circumstances.

As demonstrated, OWQM systems deliver multiple benefits and have demonstrated financial payback. The total cost per OWQM station can range from $60,000 to $120,000 depending on the type of technologies selected, availability of utilities (water, sewer, power, and communications), and desired level of sensor redundancy. CH2M HILL is the industry leader in implementation and design of OWQM stations for real-time distribution system monitoring and for broad -cope Contamination Warning System projects across the United States.

CLEAN DATA AND RELIABLE EVENT DETECTION – TURNING RESULTS FROM ONLINE SENSORS INTO INFORMATION

J. van den Broeke, F. Edthofer and A. Weingartner

s::can Messtechnik GmbH, Brigittagasse 22-24, A-1200, Vienna, Austria

1 INTRODUCTION

Most applications related to water, such as drinking water treatment, drinking water distribution or waste water treatment, are crucial for health and safety of people and the environment. For operators of water related facilities, it should therefore be a natural interest to know as well as possible the actual status of the installation and the processes taking place. To perform monitoring and control in an optimal way requires a huge number of measurement values that are continuously refreshed. Until recently, the only way to get reliable online data was by the use of hugely expensive chemical analyzers. As a result, real time measurements have remained limited to a small number of the most basic parameters. Over the last decade, more and more affordable online sensors have become available, leading to ever increasing acceptance of online water quality monitoring.[1,2] The use of such instruments allows the measurement of concentrations of substances present in the water as well as the evaluation of the overall water composition ("matrix") in real time.[3] This in turn allows control mechanisms that are optimized for and respond to the actual process conditions,[4,5] instead of settings based on less than ideal, outdated information or even purely based on operator instincts. Not only can the system be optimized, in this way it is also possible to detect failure of processes timely, and thus reduce or completely prevent damage that would be caused by this.

Because the amount of available data has increased dramatically, it is not possible to manually verify whether all these data are plausible and of good quality. For automatic interpretation to work, the data in themselves must be of high quality. This will only be possible when procedures for automatic data validation and station sensor management are used. Without validation, all results will be fed to the data interpretation system (either alarm system or process control), and faulty results will be evaluated and will be awarded the same weight as good results (Figure 1).

The need for sensor and station management is directly linked to automatic data validation. The validation aims to detect incorrect results and prevents these from being fed to the data evaluation and interpretation processes (Figure 2). However, not all sensor issues that can occur can be detected and / or identified by validation alone. In many cases, it is necessary to directly include information on activities on the station. For example, calibration of an instrument can lead to a change in the results from that instrument but a validation algorithm will need to be informed about the fact that the instrument was calibrated, as the resulting change could be due to any number of causes, not just calibration.

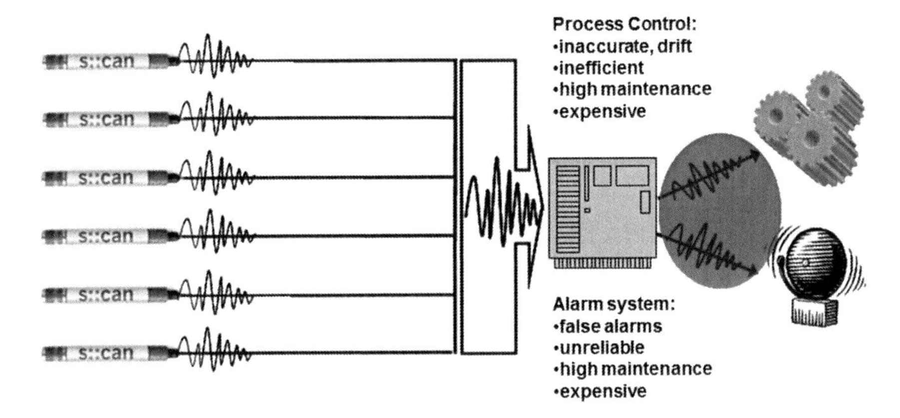

Figure 1 *The effect of measurement noise on process control algorithms and alarm systems*

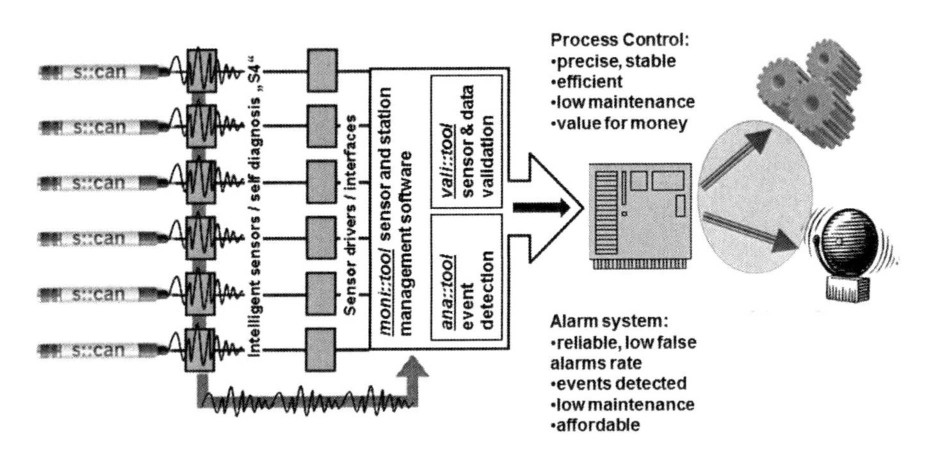

Figure 2 *Process control and alarm systems including sensor management and data validation. The depicted set-up represents the one developed here*

2 A MODULAR APPROACH TO ENSURE RELIABLE DATA AND CONTROL

The goal of this paper is to present a modular approach that looks at the whole system of user, hardware and software in order to improve the quality and availability of water quality data. The concept looks at the following aspects, each aspect being represented in the software by a separate module:

- Module 1: Sensor- and Station management
- Module 2: Data validation
- Module 3: Event detection

The separate modules are described in the following sections.

2.1 Sensor and Station Management

The most important task of this module is to provide transparency. The sensor and station management software will make available to the user all the information necessary for operation of the measuring instruments. All events and actions on the station, either concerning the entire station or those concerning single instruments (such as maintenance, calibrations, configuration, etc.) are logged and stored in a database on the station. All the data is available for review to the operator but is also available for the validation software (see section 2.2) where it will aid in evaluation of the quality and reliability of measurement results.

An important feature of the sensor and station management module is to enforce many procedures, such as entering of data on critical points as well as providing menu driven procedures such as calibrations, i.e. the user will need to enter specific information before he can proceed. When the software recognizes malfunctions (e.g. communication lost, signal out of range, membrane life expired) automatically, it actively requests information from the operator. This means that documentation of critical manipulations, from user identification and authorization to logbook keeping, is not left to the individual users and their precision in documentation, but is determined by the system itself. Because comparable systems for quality management are not yet in use, it regularly happens that measurement results cannot be interpreted as a result of very basic but untraceable causes.

2.2 Clean Data and Data Validation

The data validation module has the task to automatically detect, mark and (optionally) correct untrustworthy data. This evaluation will provide information on the functioning of individual measurements / sensors in the system. This information is then utilized in various ways, for example to provide the user with indications that a sensor requires maintenance as well as automatic detection of malfunctions. This automatic detection will facilitate more rapid detection of problems with instruments and when action is taken using the cues from the software the availability of the sensors as well as the data quality will be increased. Furthermore, through the marking of questionable results the subsequent data processing knows which results to use and which to ignore or allocate a lower importance. The use of automatically corrected results is available mainly for process control, where loss of signal can lead to incorrect settings in the process. Correction allows the controls to continue to work properly.

For the validation a number of simple but robust statistical methods have been applied, which analyze the data streams. The data is screened for four different characteristics:

2.2.1 Outliers. An outlier is an observation that does not fit to a model that is appropriate for the majority of observations.[6] To detect such observations, a smoothed curve is projected on top of the last measurement values. Using the smoothed curve, it is possible to estimate the deviation of the actual reading from the previous results. This difference is then compared to the differences between the previous measurement points and the smoothed curve. Using the information from the previous differences, a tolerance band is calculated, which follows the changes in the measurement values and which becomes narrower with higher measurement precision. The tolerance band is centred on the smoothed curve and its local width is equal to the average deviation of the measurements from the smoothed curve multiplied by a tolerance factor which is usually in the range 3 to 5. When a value lies outside of this tolerance band it is classified as outlier (Figure 3). By definition, outliers are single events or events with a very short duration.

Therefore the algorithm assumes that more than 2 subsequent measurements that are classified as outliers indicate a real change. In such a case only the first value is marked as outlier, the other data points are treated as an event (see also section 2.3).

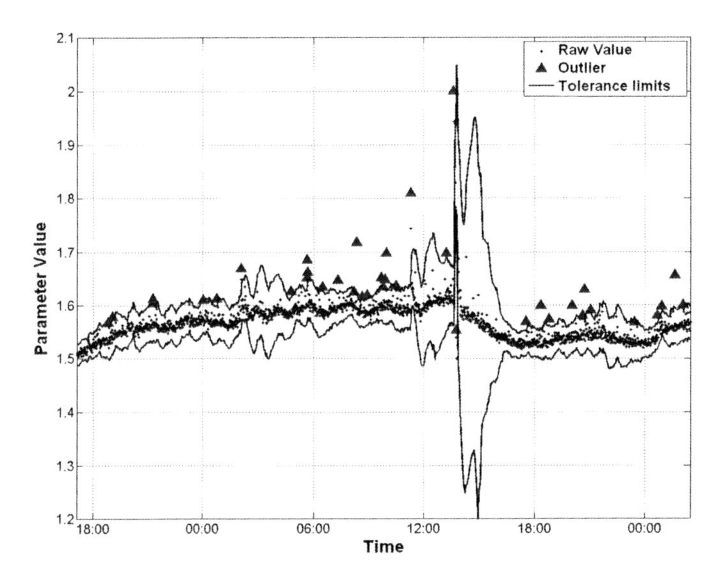

Figure 3 *Outlier detection in the data validation module*

2.2.2 Discontinuous Measurements. In case of a discontinuous measurement series, the entire series shifts by a discrete value. This is something that for example can happen when a sensor is calibrated or when the water supply to the sensor is disrupted and it is measuring in air. In order to detect such discontinuities the results are screened for intervals where the average of the measurement values before a certain time differs significantly from the average of the measurement values after this point in time. The significance of the change in average is determined by measuring the signal to noise level of the results before and after the jump and verifying that the change is an order of magnitude larger than the noise level. The change in the average constitutes a "jump height" and using the average noise level a tolerance interval is calculated. If the height of the jump is larger than the tolerance interval, then a discontinuity in the measurements will be recorded (Figure 4).

2.2.3 Noise. Noise is a measure for the random scattering of measurement results around a smoothed curve produced by using a moving average function on the same measurements. Such noise comes from instrument noise as well as slight fluctuations in concentration or activity of the measured parameters. In this data treatment no distinction is made between these two different types of noise. In the instrumentation for physical and chemical analysis, the instrument noise is typically higher than the noise induced by the dynamics of the measured parameters.

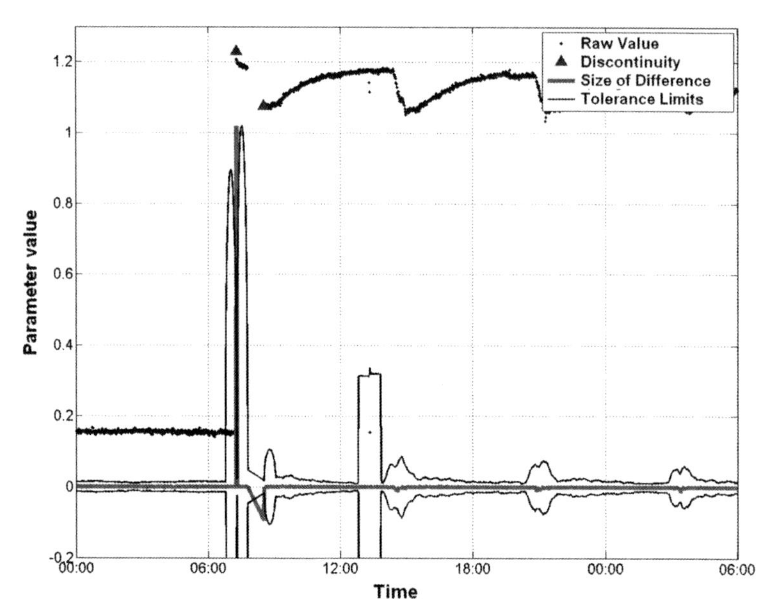

Figure 4 *Analysis for discontinuities in data validation module*

In order to calculate the noise level, the average deviation of measurement results from this smoothed curve is calculated. This noise level is compared to upper and lower limits that are known to correspond with a particular sensor or application. In case the noise level is outside of these boundaries the measurement is marked as noisy (Figure 5). This means that an increase in noise is a trigger for rejecting data in the validation, which is most likely caused by poor functioning of the sensor evaluated. It also means, that a reduction in noise to practically zero indicates that something is not working properly. Concentrations in water always vary slightly and an instrument also always produces measurement noise. The absence of noise can indicate that the signal to the instrument has been lost or that the instrument is no longer submersed in/supplied with water. Therefore this will be used to trigger a change of the system status.

2.2.4 Drift. Drift is a long term continuous increase or decrease of the readings from a measurement device. To detect drift the readings are modelled using the Holt-Winters method. This produces a slope component and when this is significantly higher or lower than zero, a drift is detected.[7,8] The time window for drift detection is significantly longer than that used for the other data validation tools; whereas tools 1 – 3 respond within minutes to changes in data, drift is only detected after increase or decrease of the values in a data stream is recognized over a period of several days.

The statistical significance of the detected drift is then evaluated against the variations in the results and only drift substantially larger than the variations is considered as drift. Drift is the most difficult of the four validation parameters to analyse without human input. It is very difficult to distinguish long term changes in the water (for example seasonal effects on surface water, e.g. temperature or chlorophyll levels) from instrument drift.

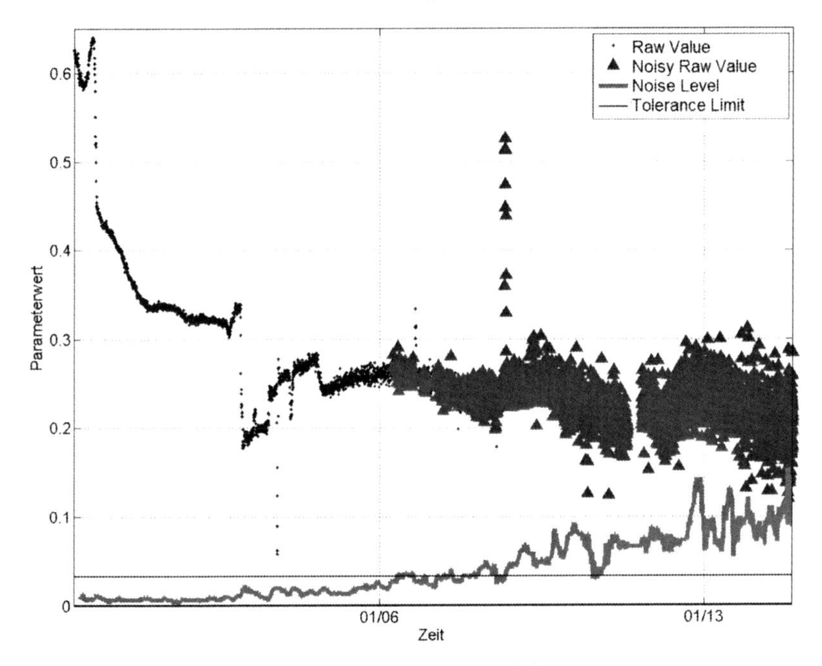

Figure 5 *Noise detection in the data validation module*

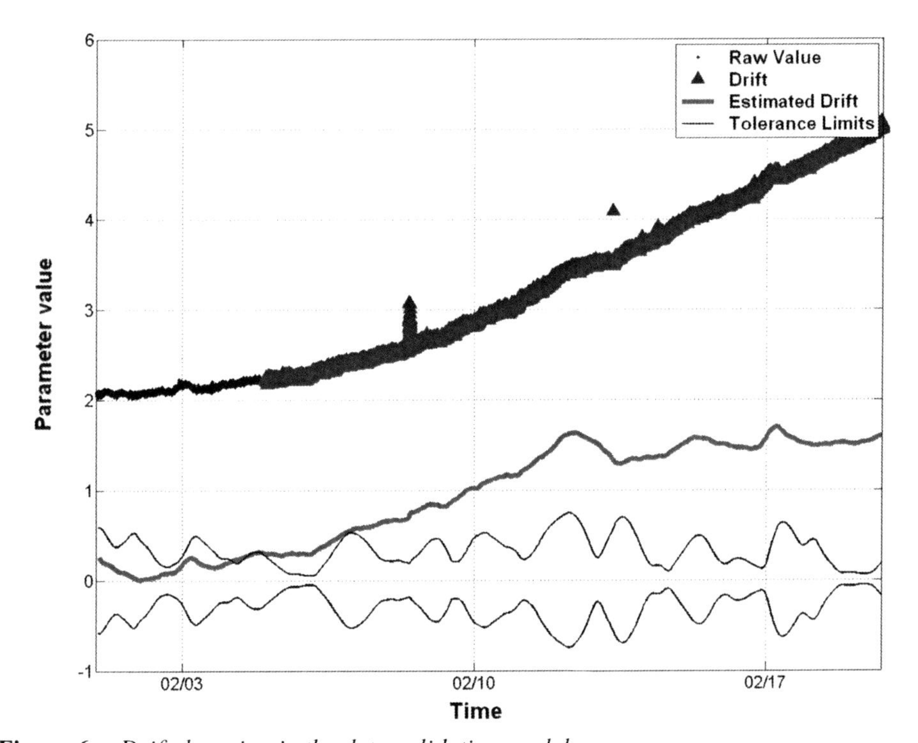

Figure 6 *Drift detection in the data validation module*

2.3 Reliable Alarming – Event Detection

Only when data quality has been ensured and when information on system status, service and maintenance activities has been brought into the equation, it is possible to perform effective event detection. The software module for event detection that is introduced here evaluates measurement data that have been cleaned by the validation module, and determines the normality of these data and triggers an alarm when a significant deviation from normality is detected. It has been optimized for the utilization of multi-dimensional spectral data, but will work just as well with single or multiple one-dimensional inputs from conventional sensors. However, the integration of spectral data provides a much more complete picture of water quality than can be obtained through single parameters, as the latter only provide a simplified projection of the multidimensional water quality onto a small number of simple parameters.[3] For example a natural change in water quality can have the same effect on pH as a toxic contamination, which should trigger an alarm. Single parameters have no way to distinguish between such causes for a change. The combination of multiple parameters into the event detection system and especially the use of more advanced tools such as spectrometric alarms, substantially improves the resolution of changes into alarms and normal changes. By calculating derivative spectra over time and space it is possible to provide further information that can be used as a basis for event detection.

The methods for the calculation of alarms in this approach consist of four different types:

2.3.1 Static Alarm. The static alarm is the most basic alarm functionality. It detects whether a single value or a group of values from a single parameter lie with the normal range of a parameter. This is done by verifying that the value does not lie above or below static upper and lower limits. In case the readings leave the limits, the system produces an alarm. The range where a value is considered to be normal is defined during the configuration and run-in period of the alarm system.

2.3.2 Dynamic Alarm. The dynamic alarm detects whether a value of a group of values shows a sudden change. Similar to the outlier detection described in section 2.2 a smoothed curve and a tolerance band are calculated. If the current measurement value is outside the limits of the tolerance band, a possible sudden change in measurement results has been detected. The selection rule used to confirm a sudden change is complementary to the rule used to mark a result as an outlier: a single value outside of the tolerance band limits is considered an outlier, multiple values outside the boundaries are considered to result from a sudden change. Such a change triggers the dynamic alarm (Figure 7).

2.3.3 Pattern Recognition. This alarm functionality uses the fact that, in case of normality in the water matrix, specific correlations between parameter values exist. This concept is best shown using a 2-dimensional example, as given in Figure 8 – in practice more dimensions are used; a series of measurements is defined as reference points. This series should present an as complete as possible image of the normal state(s) of the system being monitored. Every parameter (including every wavelength in a UV/Vis absorption spectrum) constitutes one dimension in the space of measurement values. Every multi-parameter measurement thus forms a single point in this space.

As shown in Figure 8, the measurement points are not distributed homogeneously throughout space; they cluster around reference points, which represent the normal operational conditions in the system. States that represent extraordinary situations are populated less densely.

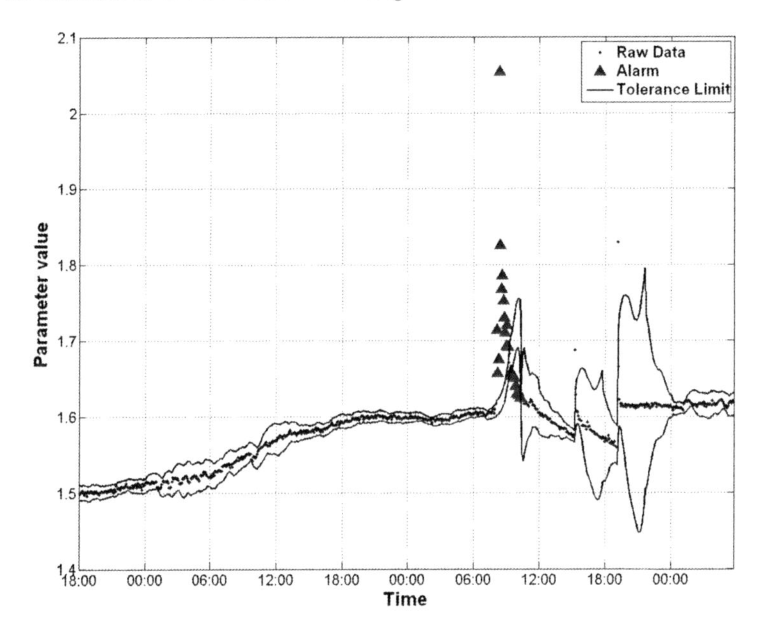

Figure 7 *Dynamic Alarm functionality as implemented in the event detection module*

The normality of incoming data points is evaluated by calculating their distance to the nearest neighbour amongst the reference points.[9] The closer a point is to a reference point, the more it resembles a known state of the system, and hence it resembles normality. A data point that lies further away from known reference points is less normal. Again, the accepted normal states are introduced in the system during a period of training. During the training the alarm limits are set based on the number of different states and the variability in the measurements. In order to eliminate effects of varying measurement ranges and scaling, all measurement values are first normalized on their standard deviation.

The above mentioned training of the system can be performed in two different modes: either a dataset with a fixed time window can be used for initial training. Subsequently, new data collected can be added so that the database expands and possibly new states of the water and different events are included in the database. The other possibility is training using a moving time window. This moving time window (e.g. a month wide) shifts constantly. Using this option the effect of long term changes, such as seasonal effects, are cancelled out and the alarm system can provide the highest sensitivity at all times. The disadvantage of this system lies in the fact that alarms that have occurred in the past are removed from the training dataset once they are outside of the time window. If these events were normal operational events and no real alarms, they will again trigger an alarm if they occur, because the system will not recognize them any longer.

Figure 8 shows three different data points and their distance from normality as well as the resulting alarm value in a trained system.

2.3.4 Spectral Alarms. This alarm functionality is only applicable when at least one spectrophotometer probe is installed. If this is the case, this module analyses the UV/Vis spectra and checks for changes in the shape of the spectrum over time. A spectrum representing normality is defined during the training of the tool. Subsequently, the deviation from normality is assessed on the 1st derivative of the spectrum. The first derivative is used as it removes practically all variability due to turbidity.[10] This results in a focus of this analysis on dissolved substances. Deviations are then determined by comparing the values at each wavelength of the spectrum with the average value and the standard deviation in that value as acquired from the training dataset. If the sum of the deviations at any wavelengths is larger than the threshold value, this can be used to trigger an alarm. The algorithm will then indicate the strength of the deviation and the region of the spectrum where the deviation occurred.

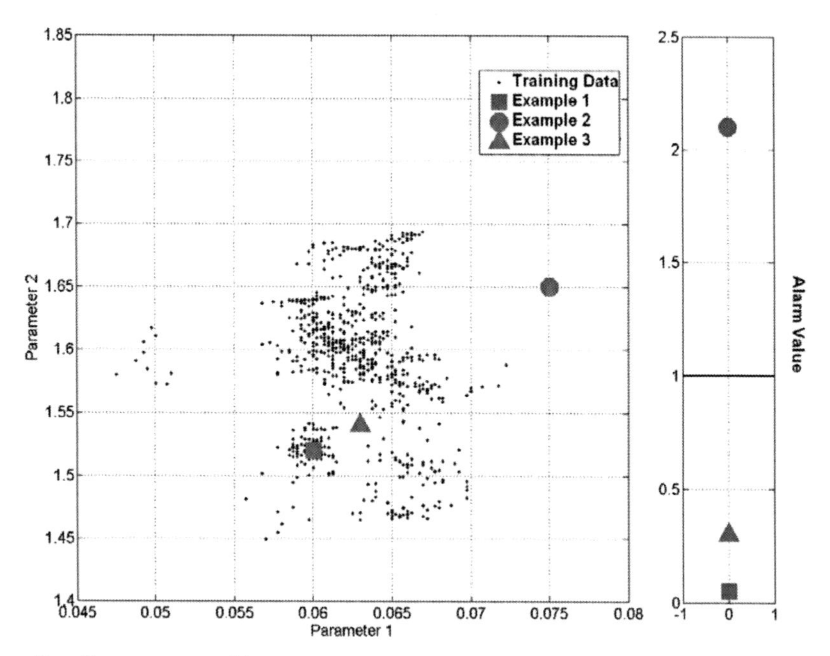

Figure 8 *Pattern recognition*

2.3.5 Cumulative Alarm. Finally the results of all individual alarm algorithms are combined into a cumulative alarm. This cumulative alarm algorithm produces the final alarm value; the alarm functionality produces only one single alarm value. This value can be triggered by any alarm individually or by a combination of alarm values, each of which might not necessarily trigger an alarm on its own. The origins of the alarm, i.e. which parameters are responsible, can be established through status information that can be accessed via logbook functionalities. However, the single alarm value is the basis for any actions that might be triggered by an alarm status, e.g. switching of valves, sending of email/SMS messages ...). The background information is available when requested but is not displayed so that it does not overload the operator with data.

The weight of individual alarm parameters is established during the training of the system. The weights allocated are determined by the number of false alarms and the sensitivity of the parameters to changes in the matrix. The results of the default alarm settings for a parameter are multiplied by a factor, which corresponds to its weight in the composite alarm. The factor is optimized in such a way that the number of false alarms is minimized. An acceptable false alarm rate of 1 per defined time window, for example month, is defined and can be adjusted by the user. The weight follows from this optimization.

This ensures that the parameters that have the highest quality (measurement noise does not dilute the information) and with the highest significance (redundant parameters have a low weight) are used. This weight of the parameters can only be properly assigned when at least one test event is available for validation of the alarm settings. This can either be a real event that was captured by the monitoring system. Alternatively, an event can be artificially generated using mathematical tools. This allows the validation of the training in case very stable water quality is monitored, and no real events are available for training.

3 EXAMPLE AND SUMMARY

All the algorithms as well as the sensor and station management modules were developed using real measurement data as well as user experience. The basic functionalities of the algorithms were evaluated using data previously obtained from various applications. The operation and integration of the three software modules is demonstrated here using a single example from a water treatment plant.

All data used in this example were collected using an s::can spectro::lyser™ submersible UV/Vis spectrometer probe.[3] Evaluated were the raw spectral data as well as parameters calculated from these absorbance spectra, such as sum organics, dissolved organics, nitrate and turbidity. Assessment of the data as well as information from the plant operators indicated that during the period evaluated here no events occurred that ought to trigger an alarm. In order to test the reaction of the event detection module to contaminants, it was decided to artificially add contaminants to the data. This was done by adding the spectrum of various concentrations of benzene (0 – 30 mg/L) to the spectral dataset and then recalculating the parameter results. The linearity between concentration and absorption (Lambert-Beer Law) allows this type of addition. Figure 9 summarizes the results of this test.

The starts of the artificial spikes are represented by the vertical dashed lines in the graph. In the middle field of the graph the measurement results of a single parameter are shown, and overlaid on these the results of the data validation module are displayed. In the lower third of the figure, the response of the individual validation methods is presented, with dots indicating that the module has estimated that a measurement results is classified as outlier, noisy, discontinuous or subject to drift. Included as well is the evaluation of the sensor and maintenance module, which in this case assesses whether the instrument requires maintenance, in this case the maintenance requires cleaning of the optical windows of the instrument. This conclusion is based upon the combined display of increased noise as well as drift, which is indicative for fouling of the windows of the instrument. It was later retraced to deactivation of the automatic sensor cleaning, which fits the observed symptoms.

The upper part of the graph depicts the alarm value generated by the event detection module, with an alarm value above 1 representing an alarm state. The first four simulated contamination events are unambiguously recognized by the system, despite the fact that the fourth spiked events is superposed on a real event that had not been reported by the operators of the plant. This underlying event could be traced back to a short period where the instrument was being supplied with water of a different composition, possible coming from a different water source. As the system was not trained on this water type, it was classified as being abnormal. The fifth contamination event is no longer detected, which can be related to the deterioration data quality which is visible in Figure 9 as increased noise on the signal and a visible upwards drift of the signal. This reduction in measurement quality is detected by the validation module, as illustrated by the increasing number of measurements being classified as noisy, outlier and / or drift affected.

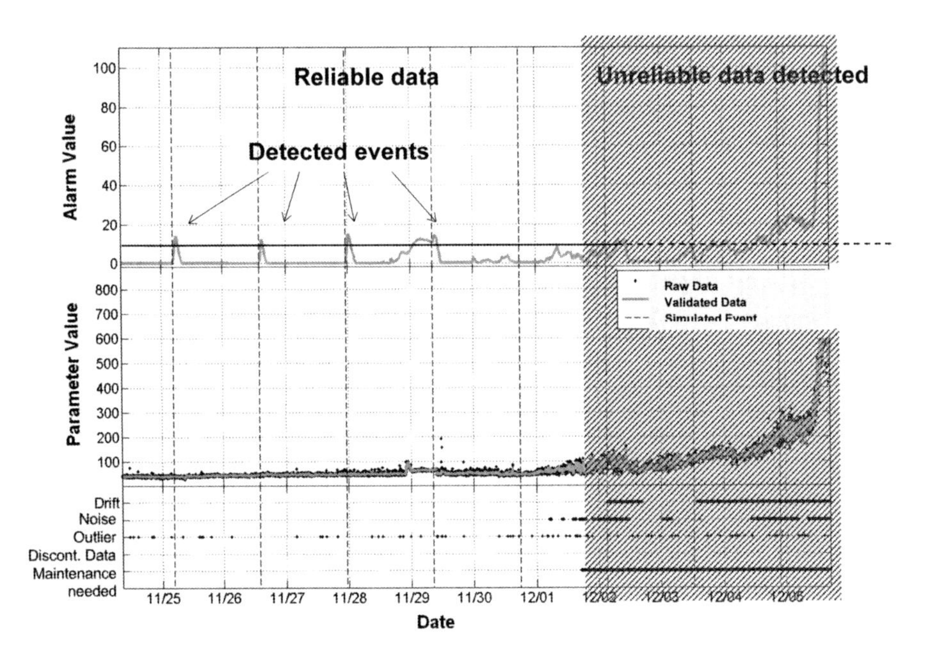

Figure 9 *Example of Data validation and Event detection on data of less than perfect quality and with deteriorating sensor performance due to fouling of the sensor*

In summary, a modular system for a reliable online monitoring and alarm system has been described. The first module that forms the basis of the system is sensor and station management. It ensures the reliable operation of monitoring stations and therefore minimizes the source of erroneous measurements at the root.

Nevertheless, measurement and installation errors cannot be fully excluded. The introduced self adapting data validation module assesses the reliability of measured data, and requires only a minimal amount of information about the monitored system. It can mark suspicious data, correct minor problems and provide feedback when data is too unreliable to be used. This helps prevent wrong decisions made on the basis of unreliable data.

Leveraging the high data quality ensured by these two basic modules, it is now possible to perform efficient and effective event detection. The event detection module described identifies unknown and unusual conditions and enables operators to react timely to faults in the monitored system.

References

1 AwwaRF and CRS PROAQUA, 2002. Online Monitoring for Drinking Water Utilities. Hargesheimer, E., Conio, O. and Popovicova, J. (eds.), American Water Works Association.

2 US EPA, 2005. Technologies and Techniques for Early Warning Systems to Monitor and Evaluate Drinking Water: State of the Art Review. Hasan, J. (ed.), US EPA Office Office of Science and Technology.
3 Langergraber, G., Weingartner, A., Fleischmann, N., 2004. Time resolved delta spectrometry: a method to define alarm parameters from spectral data. Water Sci Technol 50(11), 13-20.
4 Trieb, C. (2005). Weniger Stickstoff, mehr Leistung. AUTlook 8-9/05, 83 - 85.
5 van den Broeke, J., Ross, P.S., Edthofer, F., van der Helm, A.W.C.; Weingartner, A., Rietveld, L.C., 2009. The versatility of Online UV/Vis spectroscopy – an Overview. Techneau 2009: safe drinking water from source to tap: state-of-art & perspectives. van den Hoven, T.; Kazner, C. (Eds.), IWA Publishing, London, UK, Chapter 14 / 173 – 183.
6 A. C. Harvey (1989) Forecasting, structural time series models and the Kalman filter, Press Syndicate of the University of Cambridge.
7 C. C. Holt (1957) Forecasting seasonals and trends by exponentially weighted moving averages, ONR Research Memorandum, Carnigie Institute 52.
8 P. R. Winters (1960) Forecasting sales by exponentially weighted moving averages, Management Science 6, 324–342.
9 Cover, T., Hart, P. (1967) Nearest neighbor pattern classification IEEE Transactions on Information Theory, Vol. 13, No. 1, 21-27.
10 Langergraber, G., van den Broeke, J., Lettl, W., Weingartner, A., 2006. Real-time detection of possible harmful events using UV/Vis spectrometry, Spectroscopy Europe 18(4), 19-22.

IS IT REAL OR ISN'T IT? ADDRESSING EARLY WARNING SYSTEM ALARMS

D. Kroll[1]

[1]Hach Homeland Security Technologies, Hach World Headquarters, 5600 Lindbergh Drive Loveland, Colorado USA 80539.

1 INTRODUCTION

The concept of utilizing the measurement of multiple bulk parameters to recognize and identify water quality excursions is rapidly gaining ground as the method of choice for on-line water quality monitoring. These monitoring systems may or may not be equipped with event detection software to help in interpreting signals that indicate potential water quality anomalies. All of these monitoring systems rely upon basic water quality monitoring technology and instrumentation to provide the baseline data that is utilized in the determination of water quality excursions. With or without software, all of these systems have some potential to generate false alarms. No one wants to spend all of their resources responding to false or unimportant alarms. It is important for operators to understand that all such early warning systems are inferential in nature.

Inferential systems are monitoring systems that do not have sensors for the specific compounds of interest, but rather they rely upon a number of surrogate indicators, changes in which denote or infer the presence of the compound in question. Most drinking water early warning systems are inferential in nature. With all inferential systems, certain questions need to be answered in the shortest amount of time possible. Is the alarm real? Is the root cause in the water, the sampling system, or the measurement system? The answer to these questions is crucial to the timely implementation of response procedures.

Addressing and eliminating possible sources of error in a systematic and focused manner is the best practice in such situations to quickly arrive at the appropriate answer to these questions. Some of the critical factors to be addressed include operator interpretation of signal patterns (the duration, magnitude and shape of an alarm signal), verification of instrumental readings, knowledge of external factors and other procedures. These concepts can help in interpretation and verification of these signals. Simple rules and procedures including checklists are suggested along with things to consider when alarms happen. Following these simple guidelines, regardless of the type of instrumentation or software being utilized, will help reduce the time and effort expended in responding to alarms of all types.

2 WARNING SYSTEMS AND ALGORITHMS

The purpose of drinking water early warning systems that rely upon multi-parameter monitoring is to trigger an alarm if water quality deviations on the sensors become excessive. In some cases these software programs are capable of analyzing these deviations to see if they match an agent fingerprint from a library. It is important to understand that these sorts of systems are only capable of a presumptive classification. Any match is only indicating that the water quality sensors readings have changed in the same manner that they would be expected to if the agent in question were indeed present.

In other words the presence of the agent is being inferred from the changes in sensor readings. The actual agent may or may not be present. Occasionally more than one agent may be indicated. This is simply a signal to investigate further and can act as a guideline to that investigation. If the system were indicating cyanide might be present, it would not be wise to start an investigation on the presence of heavy metals as a first step.

3 TRIGGER SHAPE AND DURATION

There is a lot of valuable information that can be gleaned from paying attention to the duration and the shape of alarm signals. Spikes or alarms of a few minutes duration are of less concern because they affect only a small quantity of water. (Figure 1). A change that is continuous and persistent is of more concern due to the large volume of water affected. (Figure 2.)

An actual agent will usually present a characteristic rise time and a plateau of stabilization. Then a drop off will occur when the contaminated water has moved past the sensors. (Figure 3). It is possible that this kind of pattern will produce classification of different agents on the signal rise, plateau, and fall. This is caused by differences in sensor response times. Classifications on the rise and fall are not as reliable as those from the plateau of the response. This should be taken into account during the interpretation of the alarm.

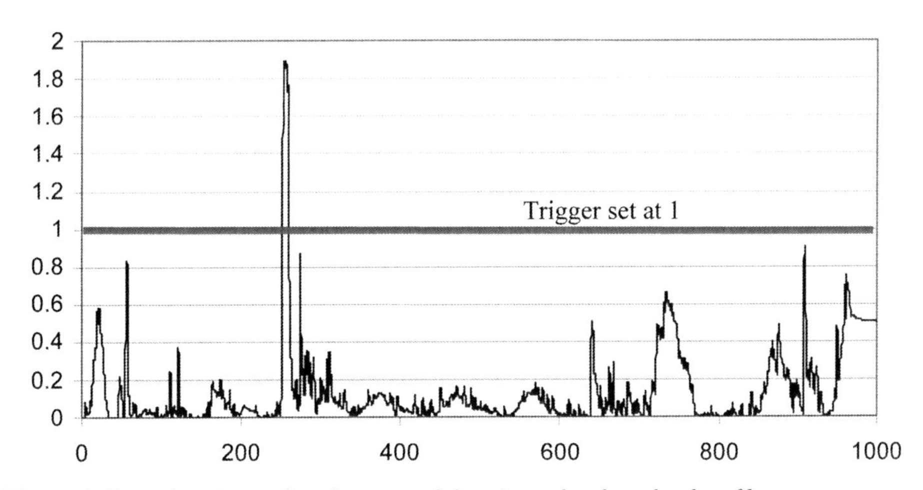

Figure 1 *Short duration spikes that exceed the trigger level tend to be of less concern.*

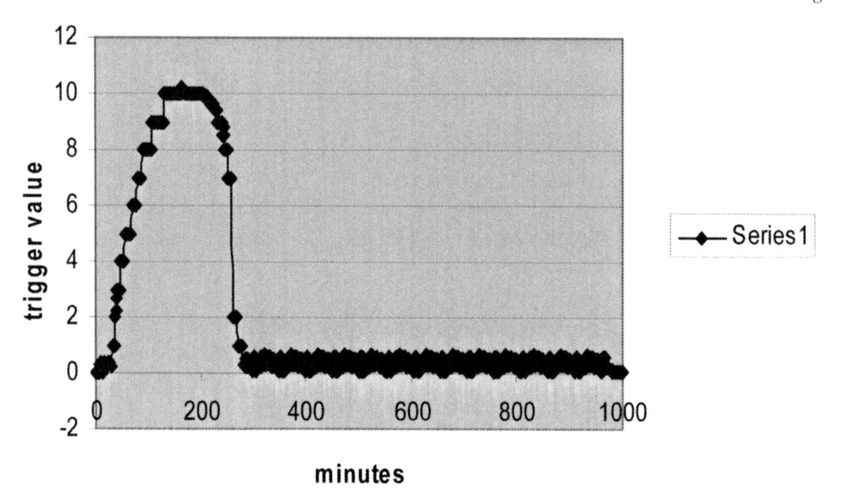

Figure 2 *Persistent changes that occur for a longer period of time are more of a threat.*

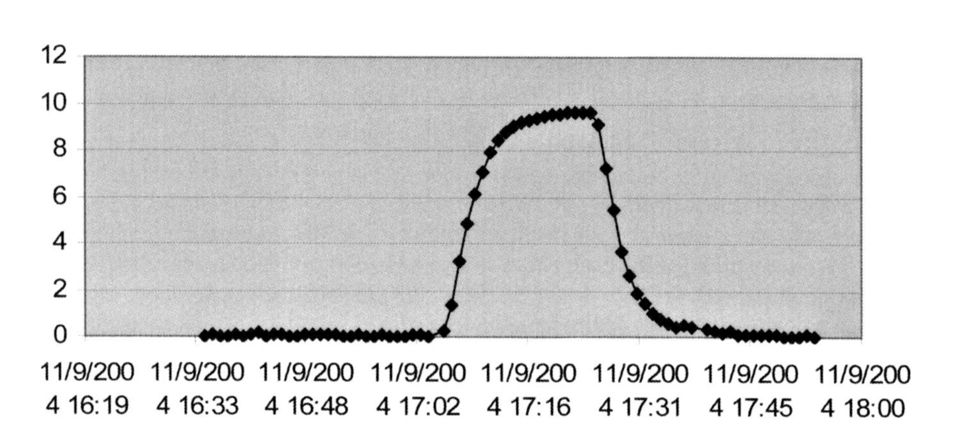

Figure 3 *An actual agent will* <u>*usually*</u> *present a characteristic rise and a plateau of stabilization. Then a drop off will occur when the contaminated water has passes.*

4 RESPONDING TO UNKNOWN ALARMS

The USEPA has performed extensive research in the area of response protocols and has developed a number of useful tools for formulating response plans. Chief among these is a manual entitled "Response Protocol Toolbox: Planning for and Responding to Drinking Water Contamination Threats and Incidents." When utilizing this tool or other response protocols, when the initial warning is from on-line water quality monitoring systems, there are a number of considerations that should be kept in mind.

When a trigger occurs (water quality readings have exceeded a threshold) there could be any number of causes, most of them of no great consequence to the end consumer.

Some analysis must be done to understand the problem and qualify the result. Is the alarm real? Is the root cause in the water, the sampling system, or the measurement system?

4.1 The Sensor Set

The first concern with a validation of an alarm is the sensor set. There are a number of things that should be considered.

- Are there any sensor alarms on the instruments that might be the cause of the trigger alarm such as a frozen sensor?
- Are all sensors functioning normally? Examine the graphs of the sensor values prior to the alarm. Is there anything unusual?
- Have the sensors been properly calibrated? Do the sensors give similar results to those obtained with independent verification techniques such as handheld pH meters etc? There are a number of useful field kits available for evaluating and verifying sensors.
- Have the sensors been properly maintained?
- Have any reagents run out? Running out of reagents on some instruments can dramatically shift readings. Those shifts can trigger an alarm.
- Have flow rates to the instruments changed? If available it is a good idea to check pressure readings to see if there have been any unusual pressure changes that might affect flows.

4.2 The Sampling System

If all of the sensors are operating without problems, then the sampling systems should be checked.

- Are there any blockages or leaks in the sampling system?
- Has the sampling system been shut off for some reason?
- Are there air bubbles in the sample line?
- Has the system been tampered with or vandalized?

4.3 The Algorithm and computer systems

- Is there a loss of communications?
- Are there sensor alarms (Hi, Low, Frozen)?
- Low-pressure alarms from any of the instruments?
- Are there any sensor diagnostic messages?

If these systems are all in order, then it is likely that the alarm is caused by a real change in water quality. A key question then becomes: Is the cause attributable to known operations or is it something unknown?

5 IDENTIFYING CAUSES OF WATER QUALITY CHANGES

When it has been determined that an actual change in water quality has occurred, data needs to be gathered regarding the operations upstream of the sampling point to see if there is any rational explanation for the change in the water quality parameters.

- Are there unusual weather conditions?
- Has work been done on or near water mains?

- Are there changes in treatment plant operations?
- Has there been a switch in source waters or a known change in these sources' quality?
- Are other monitoring sites responding in a similar manner and, is the response distribution such that it could indicate a change in the source water?
- Are different treatment chemicals being used?
- Is there maintenance occurring at the treatment plant or in the distribution system.
- Are there unusual water demands? (Major fire fighting for example.)
- A water main break? Hydrant flushing?
- Are control and feed systems for pH, chlorine, fluoride, ammonia or other treatment chemicals functioning normally?

These are just some of the possible causes that should be considered when investigating an alarm. Some early warning systems allow the operator to associate a name and an alarm priority with patterns produced by such events. The names and priorities cannot be logically assigned until the root cause of the event has been found. Once the root cause is known, these events should be named so that a recurrence can be classified quickly.

A real world event exemplifies this concept. A multi-parameter water panel coupled with a computer-based algorithm was installed at a location in a major east coast city just down stream from a water storage tank. The event monitor recoded a regular alarm. (Figure 4).

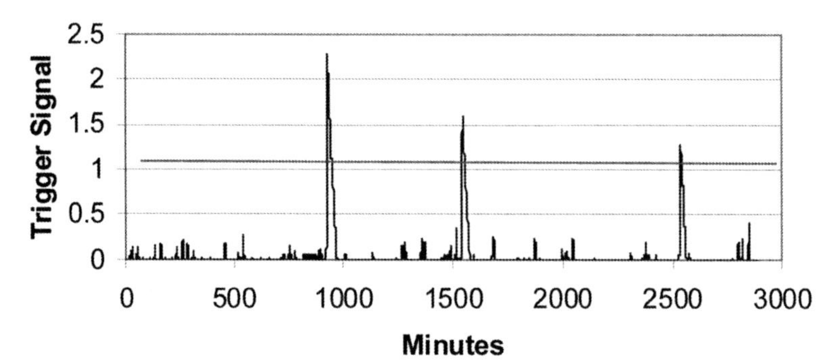

Figure 4 *Regular alarm triggers.*

Careful evaluation of the baseline parameter data showed that a chlorine upset was triggering the alarms. The chlorine levels would gradually rise over time and then suddenly drop. It was this sudden drop that was triggering the alarm. A secondary water source for the tank has higher chlorine residual than its primary source. When the secondary source was used at times of peak demand to top off the tank, the chlorine level gradually creeps up. When the secondary source is turned off, due to hydrodynamic short-circuiting in the tank, the chlorine level rapidly drops back to the concentration of the primary source.

After the operators made this determination, the learning capability of this algorithm was used to name and classify the event as benign so that when it occurred again the alarm was recognized. The system would report an upset with this pattern as **Pump shut off Normal.** Information such as this can be very useful in deciding how to respond to an alarm.

6 ANALYSIS OF UNKNOWNS

Further analysis of water quality triggers, both from classified and unclassified events, is a key component of the alarm response protocol. Many event detection systems are designed to be equipped with an automatic sampler that will draw a representative sample of the water when triggered to do so by an exceedence of the threshold level adequate to trigger an alarm. These samples can be used to perform further forensic analysis to determine the cause of an event. The utility should have adequate measures in place for analyzing samples that could be toxic or infectious.

If the system presents a classification of a likely agent, this classification should be treated as tentative until verified by further testing. As was previously stated, classification of an agent by this type of system offers a valuable first pass at determining a cause of a water quality upset and can be used to direct further forensic analysis. Tests should be tailored to address the class of agent being presented by the match.

These matches are not necessarily exact. For example, a match to an agent such as ethoprophos may not necessarily be ethoprophos but it could be anyone of a number of organophosphates with profiles similar to ethoprophos that are not in the matching library. It would be a good idea in this case to begin testing to verify the class of compounds (organophosphates) and get more specific as testing continues.

The USEAPA offers guidance on forensic testing in their emergency response protocols. There are a number of ready to use kits designed for core field testing and advanced field-testing as defined by the USEPA. High consequence actions, such as alerting customers or shutting down water supplies, should not be taken until verification of the results form the early warning systems have been performed.

Just because an alarm is not classified by an event detection system with a match to a known agent does not mean that the alarm is benign. There are many thousands of possible contaminants that could be used in an attack or accidentally find their way into the water supply. Classification algorithms usually only contain a small subset of some of the most dangerous and likely compounds. Many others are not contained in the libraries for these systems. That these systems have the ability to trigger on such compounds even if they are not specifically found in the library is one of the great strengths of the multi-parameter inferential method.

While no classification is given on these types of alarm, the information presented by the individual parameter sensors can be important in guiding forensic analysis. For example if an alarm occurs that is the result of changes in conductivity alone with no noticeable changes in total organic carbon levels or chlorine residual, it would make no sense to waste valuable time in doing an analysis focused on organic contaminants. Such common sense direction of testing can save valuable time in an emergency.

7 CONCLUSION

While not all inclusive, the concepts and considerations presented here should make interpreting and responding to alarms form early warning systems a more routine event. If followed with common sense these steps and procedures can result in increased reaction time and proper response to a variety of situations.

SELECTION AND PRIORITIZATION OF SUBSTANCES RELEVANT FOR INTENTIONAL DRINKING WATER CONTAMINATION

M. Lange and N. Pilz

IWW Water Centre, Justus-von-Liebig-Str. 10, D-64584 Biebesheim am Rhein, Germany, E-mail: n.pilz@iww-online.de

1 INTRODUCTION

The supply of clean drinking water is an infrastructure service vital to every developed society. It is provided almost incessantly to consumers and only when it is not available they realize how much they depend on it. This dependence on a service taken for granted by many makes it critical to society and thus an attractive target for attacks, e.g. by terrorists. Intentional contamination of drinking water is not an entirely new threat, it is known to have occurred even in antiquity.[1] In the past, the motive was often to get an advantage over an enemy in times of war. Fortunately, there has not been a war in central Europe for several decades. However, there is some concern about drinking water contamination driven by other motives. Threats of poisoning the drinking water were made by single offenders out of frustration or by extortionists. While only few attempts by such offenders have been reported, changes in the nature of international terrorism over the last years have raised concerns that the public drinking water supply may be at an increased risk from terrorist attacks.

Due to these concerns, the German Federal Ministry of Education and Research (BMBF) decided to fund a joint research project (STATuS) to develop a risk management concept to cope with drinking water contamination events. A key element of this concept is the knowledge on substances that may be used in an attack. A prioritised list of these substances and relevant properties will be the basis for experimental work in other parts of the project (e.g. to develop appropriate techniques for detection and elimination of these substances in case of an actual contamination).

2 PROCESS OF SELECTION AND PRIORITIZATION

Today, millions of chemicals are known. While for most of them only scarce information is available, there are thousands for which toxicity has been tested. To identify those posing the highest risk with regard to intentional contamination of drinking water a two-step process was developed within the frame of this R&D project. This two-step approach is shown in figure 1.

First, limits of toxicity and solubility (in water) were set as a minimum requirement for substances to be relevant for attacks. Combined with a half-life of at least one hour in water these were used to make a pre-selection of substances. In the second step, several classes were defined for toxicity, solubility and availability. Risk points were assigned to each class to give a score. The total score of a substance was finally calculated by multiplication of the individual scores in the three categories and a limit was set to divide priority substances from non-priority ones.

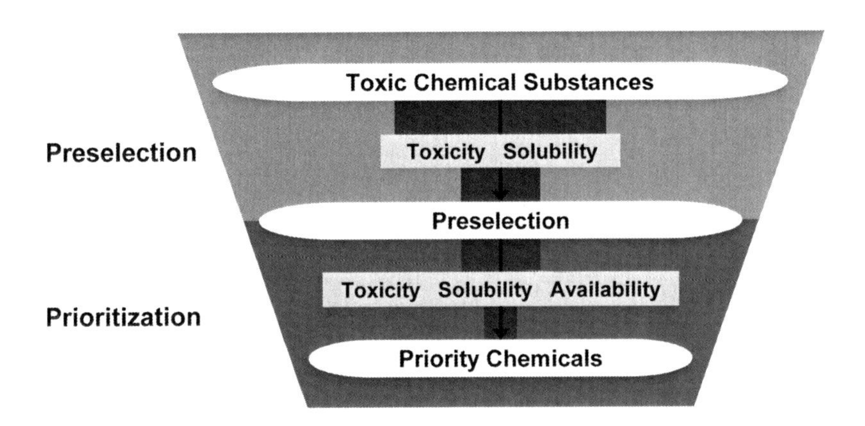

Figure 1 *Process of preselection and prioritization to identify substances posing a high risk to the public water supply when used in attacks*

2.1 Preselection

The properties usable for a preselection have to be a small subset of the total properties that characterize a substance. One the one hand they have to be sufficient to make a sound decision concerning the further consideration of the substance, on the other hand they need to be limited for reasons of practicality to ensure an effective and efficient selection. The process used to preselect substances is shown in figure 2. As the main criteria for the selection the LD_{50} as a measure of toxicity and the solubility were used. Additionally, manageability and stability against hydrolysis were taken into account. A pre-selection of about 90 chemical substances was made by application of these criteria.

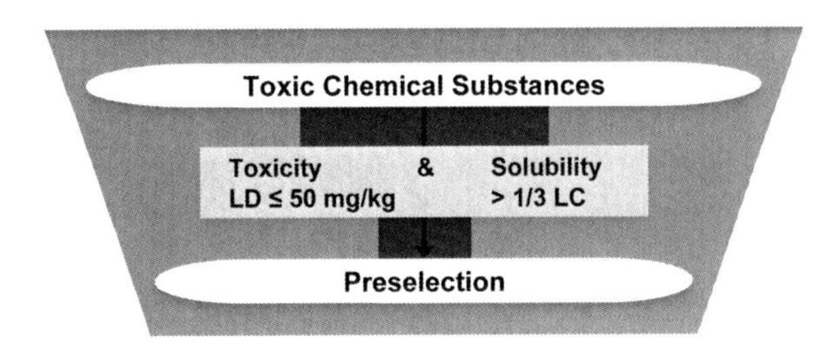

Figure 2 *Process of preselection*

2.1.1 Toxicity. Toxicity was chosen as the most important criterion since it determines the extent of anticipated health damages to consumers. In principle, substances causing only psychological effects by altering colour, taste or smell of the water may be used in attacks as well. But a psychological effect may result from almost any substance since consumers are easily scared when the media report contaminants in the drinking water. However, for a risk assessment health damages to humans were considered the main criterion.

In experimental toxicity studies different endpoints like LD_{50}, LD_{Lo}, TD_{Lo}, LOAEL or NOAEL may be reported. The one that is best defined and most commonly available is the LD_{50} (indicating the single dose lethal to 50% of the test animals). Toxicity may vary on a large scale depending on the route of administration. Possible routes are oral, inhalative and dermal uptake as well as intravenous, intraperitoneal or intramuscular injection. Because of the different efficacies of these routes of administration, oral toxicity is the only one that is appropriate for a realistic risk assessment in water contamination emergencies. Another factor that has to be considered is the variation of toxicity between different species. Since only few human toxicity data are available from accidental or intentional poisoning, data from animal studies have to be used. Here, it is crucial that the metabolism of the test species is similar to the human metabolism. Mammalian species are therefore most suitable. Based on these considerations the lowest reported oral LD_{50} for mammalian species was used as the criterion for toxicity.[2,3]

To establish a basis for the assessment of relevant LD_{50} values, a worst-case scenario was assumed where a very small storage tank of 100 m^3 is contaminated with an amount of 20 kg of a substance (this is still manageable by a single person), which would result in a concentration of 200 mg/l. Assuming an intake of 2 l and 70 kg body weight this yields a dose of 5.7 mg/kg. Thus, the selection was focused on substances with an LD_{50} below 10 mg/kg. Since toxic effects also occur at dosages below the LD_{50}, other substances were included as well when their LD_{50} was up to 50 mg/kg.

2.1.2 Solubility. An insoluble substance introduced into the drinking water in a compact form would not spread well in the supply system. As an emulsion or suspension it would spread much better, but turbidity may cause a visual warning preventing consumers from intake and causing complaints on the water quality to the utility. Additionally, the development of a stable suspension increases the effort for an offender by large. Therefore, insoluble substances are not attractive to offenders, especially since many soluble alternatives are available.

Under these premises the selection was focused on substances with sufficient solubility. Based on an intake of 2 l and 70 kg body weight a lethal concentration was calculated from the LD_{50}. Compounds were included in the selection if their solubility was at least one third of that concentration. Thus, substances were included even when their solubility was low but sufficient to cause serious health damages.

2.1.3 Manageability and stability. A substance that is to be brought into the drinking water has to be manageable by the offender with respect to the way of introduction. For example, substances that are gaseous at room temperature and not available as concentrated solutions are not suitable to contaminate large amounts of water inside a storage tank within a short time. Another factor limiting the usability is hydrolysis. A substance undergoing hydrolysis to products of low toxicity within minutes would not be able to reach consumers and thus be irrelevant. Therefore, only substances with a half-life of more than one hour in aqueous solution were included in the preselection.

2.2 Prioritization

The high number of substances posing a threat to the drinking water supply by attacks requires a prioritization model in addition to a selection scheme. It should be able to easily identify those substances with the highest risk potential. As the risk potential of a substance cannot be described by a single property, several parameters were selected for the prioritization step. For chemical substances these were toxicity, solubility and availability as shown in figure 3.

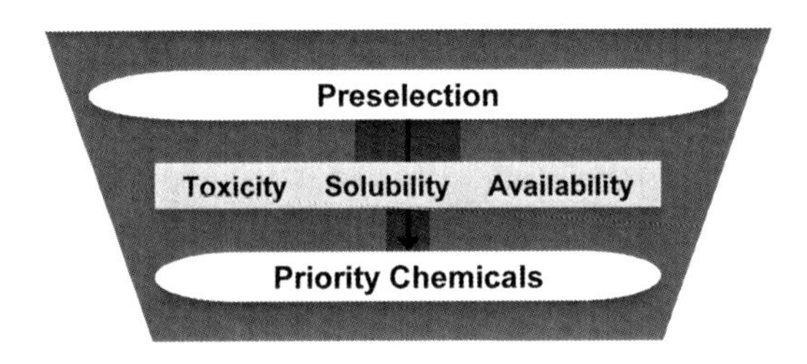

Figure 3 *Process of prioritization*

Classes of priority were defined for each parameter and risk points were assigned to each priority class. For example, when three classes were defined for a parameter, three points were used for the highest class, two points for the medium class and one point for the lowest class. To get a total score, the scores of individual parameters for a given substance were combined by multiplication. If an addition of the scores were used, a single parameter could largely influence the total score independently of the other parameters. However, the prioritization parameters were chosen such that each parameter has to reach a sufficient score to yield a significant total risk. The combination of individual scores was therefore done by multiplication as shown in equation (1). For example, a highly toxic substance with almost no availability would not pose a real risk since both parameters are necessary for that.

$$\text{Total score} = \text{Toxicity score x Solubility score x Availability score} \quad (1)$$

2.2.1 Toxicity. As discussed for the preselection of substances, oral toxicity is the central criterion to evaluate the risk potential of a substance. For the preselection a worst-case scenario was defined where a concentration of a substance is reached that would result in a dose of 5.7 mg/kg body weight for a reference person. The limit for exclusion of substances was therefore set such that the LD_{50} for mammalian species should not exceed 50 mg/kg. Starting from this premise four classes of toxicity were defined using the following principles.

- If only LD_{50} values for animals were available, the lowest available value was used.
- Where a lower LD for humans was known, it was used instead.
- In those cases where an LD range for humans was given, the mean of the LD range borders was used even if it was higher than the lowest LD_{50} for animal species.

On this basis the following classes of toxicity and respective risk points were defined for the criterion toxicity.

- **Class 1 (1 point)** Substances with an LD_{50} below 50 mg/kg but above 10 mg/kg. The limit of 10 mg/kg was chosen, since due to the variability of sensitivity between individuals and much more between different species, lethality has to be expected for some of the victims at this level, too.
- **Class 2 (2 points)** Substances with an LD_{50} from more than 2 mg/kg up to 10 mg/kg. Based on the scenario above, lethality would be expected for most of the victims at an LD_{50} of 2 mg/kg.
- **Class 3 (3 points)** Substances with an LD_{50} from 0.1 mg/kg up to 2 mg/kg. Such substances could be used to contaminate a much larger amount of water than the volume of 100 m^3 used in the worst-case estimation.
- **Class 4 (4 points)** Extremely toxic substances with an LD_{50} below 0.1 mg/kg. For these the mass of a toxic substance necessary to yield a lethal concentration is not much of a limit.

2.2.2 Solubility. Dispersion in water (which is linked to solubility) is a precondition for a substance to be suitable for contamination of drinking water. The low attractiveness of insoluble substances for offenders has already been discussed in chapter 2.1.2. Even if it were accomplished to introduce a suspension into the drinking water supply, turbidity would halt consumers from intake. Thus, an offender would aim for a toxic concentration of dissolved substances at the point of use. Since this depends on solubility as well as on toxicity, a lethal concentration was calculated for each substance using a reference person with 70 kg body weight and 2 l of water intake.

Concerning the criterion solubility the following classes with their respective risk points were defined.

- **Class 1 (1 point)** The lower limit for the exclusion of substances was set at a third of the lethal concentration. Substances whose solubility lies between this value and ten times the lethal concentration were assigned to solubility class 1.
- **Class 2 (2 points)** Substances with a solubility from ten times to below hundred times the lethal concentration were placed in the medium solubility class 2.
- **Class 3 (3 points)** When the solubility of a substance was at least hundred times the lethal concentration, it was assigned to the highest solubility class 3.

2.2.3 Availability. Substance availability was used as the third criterion, because it restrains the likelihood of a substance to be used for intentional drinking water contamination. Availability may be limited by legal restrictions, by the amount that is regularly produced for legal uses, or by the effort it takes to extract the substance from natural sources or to synthesize the compound from legal precursors.

Based on these considerations the following classes were defined for the criterion availability.

- **Class 1 (1 point)** Substances that are subject to legal restrictions on chemical weapons or illegal drugs, which are used in medicine in only very small amounts and require elaborate synthesis or isolation from natural sources as well as further compounds with limited availability from natural raw materials or low accessibility by synthesis.
- **Class 2 (2 points)** Substances that are used or have been used as pesticides but are not approved within Germany anymore, also with compounds that are easily accessible from natural sources or by simple syntheses from unsuspicious precursors.
- **Class 3 (3 points)** Substances with industrial application or availability as bulk chemicals.

2.2.4 Priority Chemicals. The threshold for substances to be considered as priority chemicals was set at a total priority score of at least 12 points. In this way about 40 priority chemicals were identified out of the about 90 chemicals resulting from the preselection.

3 SUBSTANCE PROPERTIES

A multitude of attributes can be used to depict the properties of a substance. Data that characterize a compound may be effects on living organisms, physical properties, structural constitution or fields of application. Concerning drinking water contamination not only data for risk assessment are needed, but also properties enabling the selection of appropriate detection and decontamination techniques.

3.1 General information[4,5]

Name First of all, the most common denotation was identified. For simple compounds the systematic name was used while for more complex ones trivial names were appropriate.

CAS Number Names depend on the language and may be ambiguous sometimes. CAS numbers were ascertained to use an exact reference that prevents confusion and makes database searches more reliable.

Substance class The class of a substance that is deduced from its structure is required to group similar compounds that may share the same toxicological effects. Another classification of substances can be made based on their main usages revealing similarities in availability.

3.2 Physicochemical parameters[4,6,7]

Molecular data Molecular weight, molecular formula and structural formula can be used to depict size and shape of a molecule and to reveal structural similarities between different compounds. The molecular weight is of particular interest with respect to the evaluation of membranes for treatment of contaminated water.

Solubility For most compounds experimental solubility data were available. In some cases only theoretical calculations could be used that have large uncertainties. These values should only be regarded as rough estimations since variations up to orders of magnitude may occur when different mathematical models are used for calculation.

K_{OW} An important parameter for the treatment of contaminated water is the octanol water partition coefficient. It is a measure of lipophilicity that may indicate the sorption behaviour of a substance. For organic substances experimental values were widely available. In some cases estimated values had to be used with larger accompanying uncertainties.

pK_a A parameter relevant to substances that may react as acids or bases is the pK_a. Since protonation and deprotonation can significantly change polarity and solubility of a compound, it has to be taken into account when such substances are dealt with. It was included for acids, bases and their respective salts.

Miscellaneous Other properties characterizing the state of aggregation and volatility of a substance are melting point, boiling point, vapour pressure and Henry constant. The latter being the ratio of the partial pressure of a substance in a gaseous phase to the concentration in a liquid solution. For most substances the melting point was available, for the more volatile ones also the boiling point. Vapour pressure and Henry constant could be ascertained for most of the compounds, partly being experimental values and partly calculated estimations with larger uncertainties.

3.3 Toxicity

Acute Toxicity As discussed in chapter 2.1.1 the oral LD_{50} for mammalian species is most suitable to quantify the acute toxicity of a substance with regard to effects on human health after intake of contaminated drinking water. For almost all compounds oral LD_{50} values for several species could be obtained. Only in very few cases intravenous, intramuscular, intraperitoneal and subcutaneous injection values had to be used instead.[8] Apart from quantitative data on the lethal dose the specific mode of action of a substance inside the human body and the symptoms of intoxication have to be known. They are needed to assess health risks and to choose the right medical treatment.

Approval thresholds If the distribution system has been successfully cleaned after a contamination attack, information is needed on the residual concentration of a substance that can be tolerated in the system without having adverse effects on human health. Particularly for approved pesticides and some industrial chemicals thresholds like ADI, TDI or RfD exist specifying the amount of a substance that is regarded safe for daily intake.[9,10] For some pesticides also ARfD values exist that specify the amount of a one-time dose being safe and for a few chemical warfare agents field water quality standards were available that allow for mild negative health effects.[11] All in all, for about half of the compounds it was possible to ascertain threshold values.

Ecotoxicity Not only human health may be impaired by a contamination attack but also environmental health as large amounts of contaminated water could get into rivers or sewage treatment plants. Daphnia and algae were chosen as indicators for aquatic toxicity since they are widely used as test organisms. Unfortunately, only for about half of the substances data were available from databases. Concerning algae, large variations in sensitivity to a substance may exist between different species.[12,13]

The most suitable measure of effects on the biological stage of sewage treatment would be the EC_{50} of activated sludge respiration inhibition.[14,15] Unfortunately, no data were available for the priority chemicals. The REACH regulation of the European Union requires respiration inhibition tests in some circumstances for substances produced or imported in amounts of 10 tons or more per year per company.[16] With further implementation of the regulation data will available for some industrial chemicals produced or used inside the European Union. However, most of the substances relevant for intentional drinking water contamination do not belong to that category.

4 CONCLUSIONS

A two-step process was developed and successfully applied for selection and prioritization of substances relevant for intentional drinking water contamination. First, limits of toxicity and solubility were used to make a preselection of about 90 compounds. In the second step, several classes of priority were defined for the parameters toxicity, solubility and availability. The combination of risk points assigned to the priority classes gave a total score for each substance that was used to make a ranking. In this way, the number of compounds could be narrowed down to about 40 priority chemicals.

To fully characterize these substances, an extended set of molecular, physicochemical and toxicity parameters was researched and compiled. These data will be highly valuable for future risk assessments and will be the basis for further investigation and development of appropriate techniques for detection and drinking water treatment. Furthermore, these data will feed into a dedicated database enabling fast retrieval of information on priority chemicals in case of emergency e.g., to enable a rapid implementation of appropriate countermeasures. This will be one of the key elements of the risk management concept for German drinking water suppliers.

Acknowledgements

This work was part of the joint research project STATuS, funded by the German Federal Ministry of Education and Research (BMBF) under FKZ 13N10626.

References

1 A. Mayor, Greek Fire, Poison Arrows and Scorpion bombs - Biological and Chemical Warfare in the Ancient World, Overlook Duckworth, Woodstock, 2003

2 F. Fuhrmann, A. Aigner, T. Büch, W. Legrum, C. Steffen, Toxikologie für Naturwissenschaftler - Einführung in die Theoretische und Spezielle Toxikologie, Teubner, Wiesbaden, 2006

3 U. Gerbracht, H. Spielmann, *Arch. Toxicol.*, 1998, **72**, 319-329

4 ChemIDplus, US National Library of Medicine (NLM) – Specialized Information Services (SIS) Division, Bethesda (MD), 2010

5 Römpp Online, Georg Thieme Verlag, Stuttgart, 2010

6 SRC PhysProp Database, Syracuse Research Corporation, Syracuse (NY), 2010

7 ALOGPS 2.1, VCCLAB Virtual Computational Chemistry Laboratory, Helmholtz Zentrum Muenchen, Neuherberg, 2005

8 RTECS Registry of Toxic Effects of Chemical Substances, Symyx Solutions, San Ramon (CA), 2010

9 WHO, Inventory of IPCS and other WHO pesticide evaluations and summary of toxicological evaluations performed by the Joint Meeting on Pesticide Residues (JMPR) through 2009, 10th ed., World Health Organization, Geneva, 2009

10 US EPA, Data collection for the hazardous waste identification rule - Section 15.0 Human health benchmarks, U.S., Environmental Protection Agency, Washington DC, 1999

11 US Army, Technical bulletin medical 577 – Sanitary control and surveillance of field water supplies, Headquarters, Department of the Army, Washington DC, 2005

12 ECOTOXicology Database System Version 4.0, US EPA Environmental Protection Agency, Duluth (MN), 2010

13 ETOX, Informationssystem Ökotoxikologie und Umweltqualitätsziele, UBA Umweltbundesamt, Berlin, 2010

14 OECD Guideline for Testing of Chemicals 209 "Activated Sludge, Respiration Inhibition Test", Organisation for Economic Co-operation and Development, Paris, 1984

15 Council Regulation (EC) No 440/2008 of 30 May 2008 laying down test methods pursuant to Regulation (EC) No 1907/2006 of the European Parliament and of the Council on the Registration, Evaluation, Authorisation and Restriction of Chemicals (REACH), last amendment 27/08/2009

16 Regulation (EC) No 1907/2006 of the European Parliament and of the Council of 18 December 2006 concerning the Registration, Evaluation, Authorisation and Restriction of Chemicals (REACH) [...], last amendment 02/04/2010

THE NEED FOR A JOINED UP APPROACH TO THE PROVISION, MANAGEMENT, SECURITY AND DELIVERY OF ALTERNATIVE DRINKING WATER SUPPLIES

K. Silcock

Managing Director, Water Direct, Earls Colne, Colchester, Essex, CO6 2NS, UK

1 INTRODUCTION

When the piped drinking water supply system is unusable, inadequate, or just not available, an alternative supply of wholesome water is needed. If the infrastructure becomes unavailable or unusable, due to either a failure in the system or a contamination, that alternative supply will be needed quickly. In our developed society, we take water for granted and when the taps run dry people tend to panic.

There are a number of methods used when deploying alternative drinking water supplies. The most recognised is the deployment of water bowsers to the street where customers can collect water in their own containers. Static water tanks are also often used, which are filled from water tankers. In each case, the vessel deployed will carry a 'boil before use' recommendation notice. This is to safeguard the public and the provider, should the water become contaminated or its wholesomeness deteriorates between the customer collecting it from the tank and getting it home. Some water undertakers have trialled the provision of carry home containers when deploying bowsers and tanks, although the boil notice would still be advisory.

Bottled water has therefore become an ever more popular way to provide alternative drinking water supplies, especially as it can be deployed quickly and cost effectively. The customer receives water for drinking and cooking quickly, which calms the situation, reducing the pressure on the undertaker whilst trying to restore the piped supply, and can be consumed without boiling first.

This alternative to the piped supply, whether provided in containers, bottles, tanks or tankers is an extension of the Public Water Supply and as such the security and integrity is therefore paramount; so managed processes and procedures must be in place in advance of the event to ensure that the water remains wholesome, from its source right through to when the public consumes it. In a water supply emergency, the last thing the public needs, or will be expecting, is a sub-standard or unwholesome alternative supply.

The water industry is familiar with the processes required when delivering alternative supplies in tanks and tankers. A frequent misconception still appears to be apparent with bottled water though, that once it has been filled and sealed it must be OK when the seal is first removed. Mostly this is true; but the real issue is how do you know, when the only sampling is during the bottling process?

2 THE WAY FORWARD

In the United Kingdom the Public Water Supply is governed by Law under the Water Act. Within this framework is the Security and Emergency Measures (Water & Sewerage Undertakers) Direction (the SEMD). This lays down the responsibilities of water undertakers to provide certain minimum volumes of wholesome water to the Public within 24 hours of a water supply failure and each day thereafter, until normal supplies are restored. Currently these minimum volumes are 10 litres per person per day, rising to 20 litres per person after 5 days in a major or widespread event. Water undertakers must have plans in place, which are audited by DEFRA, and be independently capable of serving specific numbers of people, allied to population served by each undertaker. These numbers range from less than 10,000 to as many as 30,000 people within the first 24 hours, beyond which regional Mutual Aid and eventually more widespread assistance may be requested. The Water Act provides a mechanism for monitoring performance and for prosecutions to be brought against water undertakers that fail to meet the requirements of the SEMD.

In the rest of Europe the regulatory framework appears to be different. Only 3 litres per person per day is required and 'best endeavours' is apparently acceptable for this provision, with no mechanism for penalties for failure. Some may say we are out of step with Europe. I would argue that the UK has set the better standard in regard to the acceptance of responsibility and focus on customer service.

The quality of the Public Water Supply in the UK falls under the Water Supply (Water Quality) Regulations 2000, as amended, with the Drinking Water Inspectorate (DWI) as regulator. When a water undertaker deploys alternative supplies, using bottles, containers, tanks and tankers, the regulator determines them to be an extension of the Public Water Supply and as such they must also conform to the same regulations. Due diligence for conformity with any form of supply must include sampling and compliance cannot be adequately demonstrated without this.

The point of compliance for water provided through the piped supply is when the water first emerges from the tap. The point of compliance for water provided in bottles and containers, as clarified in Guidance (Edition 3, September 2008) to the Water Supply Regulations, is when the water first emerges from the bottle or container. For bottled water drawn from stocks held for emergency use this could be some months or years after the sampling undertaken by producers at the time of bottling; so a regime of regular sampling and monitoring of stored bottled water is essential to demonstrate conformity.

Again, in the rest of Europe the definitions for compliance do not appear to be as clear as in the UK. The assumption that water in sealed bottles and tested within 12 hours of production will meet the wholesomeness standards at the point of delivery for example is an arguable point; especially if no diligence is exercised or samples analysed between production and deployment in an emergency. No mechanism exists in that scenario for demonstrating compliance. The quality of bottled water produced in the UK is of the highest standards. There have however been a number of UK and European bottled waters recalled due to quality issues, which have arisen or been detected after the final product was released from production. Some of these contained faecal contamination.

The reason for sampling stored bottled water is not necessarily to find problems; more importantly it is the methodology necessary to enable stability and compliance to be demonstrated against expectation and observation. The mere fact that bottled water is held in storage for emergency use enhances the diligence of being able to demonstrate quality conformity at the point of compliance; a factor not practically possible when taking bottled water straight from a previously unknown or unfamiliar production facility or even a retail outlet. Stored water will not normally be released to a consumer for at least 6 months after production due merely to proper management of stock rotation. This provides an inbuilt 'quarantine' situation within which some testing and monitoring occurs.

Whilst a great deal of discussion, reassessment and new planning has occurred since the summer floods of 2007, when 360,000 people in the UK were unable to use the piped water supply for up to 9 days, there is still much work to do in establishing common standards and levels of preparedness for when it happens next, even in a much smaller way; as it does frequently.

When a water outage affects a single water undertaker, efficient and effective deployment of an alternative supply can be achieved, as that undertaker is in complete control of its own planning, methodology, procedures, priorities and implementation. Staff are familiar with the systems and have trained and practiced. But even so there can be gaps. Under the SEMD, priority must be given to domestic customers, with due regard to the needs of the elderly and infirm. Without prior planning however, even hospitals and other health care facilities may experience delays in receiving an adequate or even appropriate alternative supply. As mentioned at the beginning, our society takes water for granted and the end user needs to be aware of what to do when an outage occurs and have plans in place to react quickly to mitigate consumption and reduce or avoid distress.

In the wider commercial world, there is no regulatory requirement for a water utility to provide a continuous piped supply 24 hours a day. There is no guarantee of an alternative supply either. Yet so many businesses rely on such a facility being available on a constant basis, with little thought given to what they would do should the supply fail; other than to call their water undertaker. Most would then hear that an alternative supply in the short term, from a regulatory responsibility standpoint, would not be possible and this could be functionally or financially devastating for some businesses. Consultation, education, planning and prior preparation are therefore essential as part of good customer service.

To sustain the infrastructure of our communities these businesses must continue to function and an often overlooked priority in receiving an alternative supply, which is directly related to this, is schools. For reasons of Health, Safety and Hygiene schools will not remain open if the piped water supply fails. Yet the disruption to our communities and the businesses that support them of a school closing unexpectedly can be huge, as parents must decamp to collect and/or care for their children at home. Prior planning for providing an appropriate alternative supply quickly to these facilities would negate this adverse reaction.

When a water supply incident affects more than one water undertaker concurrently, such as an accidental contamination, the industry has a tried and tested method of working together and supporting each other. However, there are still a number of areas where improvements could be made. These are summarised as:

1. Attitude to risk
2. Commitment to funding
3. Commitment to preparedness
4. Compatibility of equipment and resources
5. Joined up central management

1. Attitude to risk;

Some water undertakers, and many individuals within them, have never experienced a substantial water loss event and a degree of complacency may therefore exist based on probability. A major water loss may be deemed from risk assessments and inbuilt resilience of the infrastructure as low risk and this will determine the level of commitment to being prepared for such an event. But whether a large incident or small local event the principles of preparedness are the same.

2. Commitment to funding;
If risk is deemed low then why commit a high level of funding to it? Many of the incidents we are called upon to support were never anticipated in advance as being high risk, but they happened nevertheless. The only predictable facet of water supply loss is that it is entirely unpredictable.

3. Commitment to preparedness;
Incident recovery can always be achieved faster, more effectively and at less direct cost when prior planning and preparedness has been put in place. Vessels to be used in an emergency must be fully prepared, disinfected, sampled, sealed, stored and maintained ready for immediate deployment. Bottled water must be pre-produced, adequate, managed, quality checked and available for that essential first fast response and sustainment. Transportation resources must be available in the right place at the right time with this emergency response as its first priority. All of this costs money, when there is no guarantee that it may ever be needed – just like insurance. But can you risk not having insurance? You cannot make a claim if you don't pay the premiums in advance.

4. Compatibility of equipment and resources;
Both the Pitt and Water UK Reviews of the flooding events of the summer of 2007 make mention of the lack of, and the need for, compatible equipment. Water undertakers in the UK are independent commercial businesses, with their own procurement processes and strategies. This determines that differences in types of equipment purchased for individual use, or methodology of how to procure resources for emergency use, will always be evident. But when the industry acts collectively in a major or widespread event there needs to be standardisation. Water tankers are a classic example. Unless they are specifically built and dedicated for drinking water they will be spot hire food grade tankers, used to carrying any liquid foodstuff, from milk to molasses. Apart from the cleanliness and therefore fit for purpose issue, these tankers have different hoses and couplings for filling and discharging their products. Few if any are the same as those used in the water industry. None will have been adequately cleaned and disinfected for use with drinking water. Under that strategy, apart from the delay in cleaning and preparing the tanker for drinking water use, a whole plethora of hose adaptors must therefore be available – but how many and where do you keep them?

5. Joined up central management;
The UK water industry's Mutual Aid scheme, which is a method for each company to assist others in need, works very well in the spirit of co-operation. But, probably for the reasons of procurement and attitude to risk mentioned above, differences are still evident in regard to compatibility of equipment, degrees of preparedness and strategies for deployment. Not all undertakers work to the same levels of preparedness for example and will have different priorities or strategies for methods of deployment and of how and when which customers receive alternative supplies.

Within the water supply industry in the UK, and potentially across Europe, there is an urgent need for more joined up thinking and collective action with immediate access to centralised assets, equipment and resources which are dedicated for the provision of quality assured wholesome alternative water supply solutions.

But, the non-availability or inadequacy of a piped water supply is not always the responsibility of the licensed water undertaker. This may be due to bursts or contaminations within a customer's boundary, internal system failures, or simply that a wholesome supply is required where no infrastructure exists. There are many examples of this, such as;

- Construction sites where there is no access to a supply, or it is inadequate
- Welfare and hygiene for remote or transient workers who have no access to a piped supply
- Food production facilities where peak production demands exceed what is available
- Outdoor events such as festivals, concerts and camp sites, where no infrastructure exists or is inadequate for the event
- Customers' internal system failures, where provision of a fast alternative is essential for sustaining production, hygiene and occupation to keep the businesses and our economy running
- Agricultural facilities where private sources have failed or become contaminated

All of these arenas require the same standard of clean, quality assured, wholesome drinking water that is free from substances that are, or may be, harmful to human health. To ensure that, common standards of provision and diligence for compliance are needed. And this does not exist currently in the commercial market place.

Acknowledging who is responsible for the provision of wholesome water to the public and to employees is therefore also a critical factor. Is it:

- The water undertaker?
  - Yes, in whatever form the water is provided by them to the public
- The local council?
  - Yes, if they are involved in the procurement and provision of alternative supplies to the public
- The government?
  - Yes, in determining and policing the required standards for licensed undertakers
- The Environmental Health Officer?
  - Yes, in being aware of and ensuring the required standards are maintained
- The commercial provider of alternative supplies?
  - Yes, and they must be diligent at every stage
- The water source owner?
  - Yes, under the private water supply regulations
- The bottler?
  - Yes, under the bottled water regulations
- The employer?
  - Yes, when procuring an alternative supply, in whatever form, for their employees
- The charity or event organiser?
  - Yes, when providing water free to the public
- Individuals involved in the supply process?
  - Yes, all have a duty to be diligent

Everyone in the supply process is responsible in exercising all due diligence to ensure that the water is, and remains, wholesome and free from harmful elements, as far as it is within their control.

Critical issues that raise barriers to the efficient provision of a quality assured alternative drinking water supply include:

- Lack of awareness or knowledge of responsibilities
- Lack of awareness or knowledge of best practice
- Lack of funding commitment – leading to lack of planning, preparedness and dedicated resources
- Mixed views on the point of compliance and assumptions as to who is responsible, especially in regard to the use of bottled water
- Absence of appropriate logistics, which needs to be available at short notice
- Reliance on poorly prepared or non-dedicated supply chains that are not geared to respond in an emergency
- Disjoined management, leading to a lack of standardisation of equipment, processes and methodology
- 'Unqualified' or uneducated third party providers of what is a quality critical service

When it goes wrong, the public is at risk. Providers of drinking water, in all forms, at every stage, need to be made aware of and fully understand their responsibilities; and the boundaries of responsibility.

Some real examples of when things go wrong include:

- Bottled water product recalls – when normal sampling regimes at time of production have not prevented product release
- Drinking water tankered into a festival camp site which tasted of milk
- Temporary storage tanks and distribution pipework set up for an outdoor event that were not cleaned, disinfected or sampled before use, resulting in contamination of the supply and cessation of the event
- Temporary tanks set up for welfare water on a building site that were refilled by the waste collection tanker; where saving money was more important than safeguarding public health

The consequences of a lack of diligence are serious and could even be fatal.

3 RECOMMENDATIONS

What is required for a fast, reliable and sustainable provision of a quality assured drinking water supply by alternative means is a resilient system. First and foremost what is needed for resilience, abundance and sustainability is multiple, accessible and geographically disparate water abstraction locations and bottled water production facilities.

Then, assets that are dedicated for drinking water and are not used for any other purpose. Ideally this will involve common stockpiles of bottled water, containers, vessels, tankers and associated equipment; which are prepared, sampled and commonly managed; adequate, available and ready for immediate deployment anywhere, fast. Common specifications, processes and methodologies would ensure equipment compatibility, adequate levels of preparedness and clarity of quality monitoring for compliance.

Unilateral responsibility for management could be sustained by adopting a common 'best practice' system of provision, adopted by all, which would be effective on a single undertaker local event basis but can be simply replicated and escalated to service major or widespread multi utility events.

In parallel to this there must be a multilateral responsibility for quality assurance involving clear responsibility and due diligence throughout the supply chain, which will enable assurance of quality and wholesomeness at the point of delivery.

Preparation is critical and cannot be stressed enough. Proper preparation in advance of an event will cost some money, but will enable the response to an unplanned event of any scale to be undertaken in a more controlled and cost efficient way.

With bottled water continuing to grow in favour, as the preferred first response to a water loss event, the importance of proper processes increases concurrently. A resilient system for the provision of an effective and quality assured response with bottled water must include:

- Multiple production facilities for resilience and sustainability of supply
- Pre-produced stocks which are labelled and packaged appropriately
- Regular sampling and monitoring for quality assurance at the point of compliance
- Managed to ensure adequacy and sustainability
- Managed for stock rotation within best before dates
- Managed for enabling fast escalation and sustainable replenishment
- Located within a few hours travelling time of delivery points

Water Direct established such a facility at the request of a group of UK water undertakers 12 years ago. The facility, now known as the Nationwide Bottled Water Bank, is available across the UK and plans are in place for enabling fast escalation and sustainment to major events. This sits alongside an Alternative Water Container Bank, which is a similar facility for vessels and tankers.

4 CONCLUSIONS

In conclusion, drinking water provided in any alternative form to employees and to the public must be wholesome, as defined within the Water Supply (Water Quality) Regulations 2000, as amended. The provider, whoever that may be, has a responsibility to ensure that it is free from contamination and not harmful to human health – at the point of compliance; which for bottled water is when the water first emerges from the bottle. This responsibility must be understood by all providers and may be enforced by regulators.

In an emergency you need assurance that the water you provide is wholesome and immediately available in an appropriate form. Alternative supplies must therefore be prepared and fit for purpose; but they are not just for emergencies.

Already today there is a greater reliance on the availability and prompt delivery of alternative water supplies. Our threatened environment reinforces the need for planning and preparation. So the time is right therefore for a multi-agency joined up approach to agreeing standard methodology with compatible resources and robust processes for quality assurance, to ensure that the same high standards for quality delivery are met by all providers, in every arena.

Our piped water supply infrastructure is under stress from increased demand, rising costs and the pressure on maintaining wholesomeness from source to tap. Add to this system malfunction, human error and the threat of deliberate damage and I believe that within the next 10 to 20 years the logical solution for providing water for human consumption, which is only a small percentage of the volume flowing though our piped supply systems, will be direct to the consumer in bottles and containers.

So we must get the system right, now.

COMMUNICATING WITH THE PUBLIC ABOUT RISK

M. McGuinness

Scottish Water
Castle House, Dunfermline, Scotland

1 INTRODUCTION

There are few greater threats than those that affect our health or wellbeing. We all need a clear understanding about risks to our health; these demand clear, relevant, accurate and timely information before we can make decisions.
The effective communication of any risk is a pivotal part of any public health strategy. The methods used must be informative, educational and advisory to allowed informed choices to be made.
This communication is often a challenging process and can at times be extremely complicated, each situation is different and the need for clear communication of hazards and risks must be carried out to avoid confusion and anxiety.
It is important that all organisations feel themselves to be in the best position to effectively and accurately communicate risks to their customers, there needs to be an emphasis on working with the media as this is generally the major route of communications to the public.
The communication of risk is always difficult to do well; many elements, cultural, political etc. can influence how well an organisation can perform.
Risk communication takes place at times of uncertainty, true facts may be unclear, information incomplete to drive appropriate response, and that is to say such communication may have to take place during a rapidly changing situation.
This is why clear guidance in such communication is required to ensure the right message is delivered at the right time, in the right format based on best information available.
Communication is a two way street and the building of constructive partnerships with all stakeholders involved is important, such partnerships are built on trust and must be continually reviewed to ensure they remain effective.

This paper looks at:

- Risk Communication – the definition
- Risk Perception – the characteristics of risk
- Risk communication and audience – public, media, and academics
- Planning risk communication – the elements of the process, delivering the communication, the barriers to communication
- Communication skills – hearts and minds

This is based on collaborative work that was carried out in Scotland by Health Protection Scotland, NHS Scotland and other interested bodies including the BBC and Scottish Water to develop Communication Guidance which has been adopted by Scottish Water in its Public Health division to ensure clear and consistent communication of risk to our customers.
The Health Protection Advisory Group directed Health Protection Scotland to bring together a working group to develop evidence based guidance for public health risk communication.
The group developed systematic evidence based methods to identify and appraise risk communication guidance. Consensus techniques were used where evidence of best practise was lacking.
The group used a development algorithm and identified seven extant guidelines on risk communication, these were then subject to appraisal using a modification of the AGREE instrument. (Appraisal of Guidelines for Research & Evaluation)
It was decided that the development of formal evidence based guidelines with evidence standards would not be feasible and that developing peer reviewed guidance was a more credible option.

2 RISK COMMUNICATION

The US National Research Committee on Risk Perception and Communication defines risk communication as;
"An interactive process of exchange of information among individuals, groups and institutions. It involves multiple messages about the nature of the risk and other messages not strictly about risk that express concern, opinions or reactions to risk messages or to legal and institutional arrangements for risk managers"

Risk communication, by its very nature, intervenes into complex systems and many different factors must be taken into account. It is a dynamic process that is informed by a huge number of separate influences, many outside our control. Communicating one risk often reduces it, but replaces it with another – a boil restriction for example where the risk of drinking unwholesome water is replaced by the risk of scalding associated by handling boiling water.

Risk communication must:

- Have clear objectives which are continually under review, that aim to improve general understanding of the risk.
- Strengthen working relationships and promote mutual respect among all parties.
- Assist in the development of a consistent; transparent and credible decision making process.
- Ensure that all advice and information for the general public is clear and timely.
- Make clear from whom the message comes and what role that authority is playing.
- Foster public trust and confidence in risk management decisions.

The aim of risk communication is to provide the public with meaningful, relevant, accurate and timely information, a major part of this communication is that it must establish the trust of those to whom the risk is being communicated. Best practice shows that to achieve this, the communication must be empathetic and caring, transparent and open, be well planned and communicated at the earliest opportunity.

3 RISK PERCEPTION

Individuals hold different perceptions of risks, this poses a challenge to communication. Perceptions can be influenced by many factors including:

- The risk is inequitably distributed
- The risk is inescapable
- The risk comes from an unfamiliar or novel source
- The risk is man-made rather than from a natural source
- The risk may cause hidden or irreversible damage
- The risk may affect small children or pregnant women

4 RISK COMMUNICATION AUDIENCES

There are multiple audiences for any risk message. Many issues are of national or international interest, but the primary audience needs information on what to do locally and it can be easy to lose sight of the needs of those most at risk.

The public are not a homogenous group, different people have different interests and may seek information in different ways. Therefore it is always useful to identify the target audience:

- Those affected by the hazard
- Those at risk from the hazard
- People in other areas who are not at risk but who may perceive themselves to be
- People in other areas who find the issue of interest

Clarity as to the audience helps achieve clarity of message. The message for those affected may be what is being done, for those at risk it may be what steps are being taken to minimise risk. While for those who fear wrongly that they are at risk it may simply be that the incident does not affect them.

The public are of course not the only audience there is also the media, in an incident the media need to be dealt with either proactively or reactively, the public has a right to information and the media can be critical in disseminating the correct information.

Recognising the needs and perspectives of the media is essential in framing a successful overall communication strategy. Attempts to engage meaningfully with the media may also help to avoid the often inaccurate speculation that inevitably develops in an information vacuum.

Consideration should always be given to working positively in partnership even if initial involvement may seem negative or confrontational. People in the media see themselves as having an important role in informing the public and holding officials to account, particularly where there is a perception of risk and culpability.

The media will provide information to the public with or without your help so it is always in your interest to work with them to ensure that the correct message goes out.

The final community that needs to be carefully communicated with are the academic communities; here also good communication is crucial. The academic community has an input to both acute and more long standing risk communication endeavours. They can provide much needed knowledge and support, particularly where contentious debate is likely.

5 PLANNING RISK COMMUNICATION

Developing, exercising, implementing and regularly updating a risk communication policy and strategy is central to successful risk communication.

US Centre for Disease Control and Prevention states:
"No organisation should consider itself prepared to respond to a crisis if it does not have a communications plan fully integrated into its overall disaster response plan".

What makes a crisis communications plan a good one is the process used to develop the plan rather than what ends up on paper.

The questions we need to ask are:

- What information needs to be in place
- Who makes the decisions
- Who gives orders and who follows them
- What are the procedures for carrying out response initiatives
- Who will be the lead spokesperson

A crisis is not the time to be thinking about these questions. In fact it is the worst time to do so. Agreed strategies should be in place, contact lists should be in place and up to date, and all briefing materials prepared.

Planning the communication process is of the utmost importance in ensuring the correct messages are received. The communications strategy should:

- Define the issue or problem
- Identify the stakeholders and target audience
- Set detailed objectives
- Select key messages
- Engage with partners that will be involved in managing the incident and who need to contribute to key message development
- Choose the most appropriate communications channels
- Track and evaluate the impact of any communications

Providing clear leadership is always important in any situation and is paramount to successful management. In an ideal world the leader and communicator would be the same person, but these roles are often carried out by different people. Therefore it is important that there is close contact between the two and that lines of authority are agreed early on.
Staff who face the media will need appropriate training in:

- Dealing with facts
- Working with time and space constraints
- Interview situations
- Presenting information at public meetings
- Dealing with reactions to risks
- Developing trust
- What types of risk are seen as acceptable
- Risk communication strategies

When delivering risk communication messages it is important to be precise and specific, focus your language on positive options, reinforce your message by repetition and create mechanisms to enable people to remember the message.

Keep your messages:

- Simple – frightened people don't want to hear big words
- Timely – frightened people want information immediately
- Accurate – make it direct, don't grasp nuances
- Relevant – action steps and answer questions directly
- Credible – use empathy and openness
- Consistent – keep to your key messages

When choosing your messenger look for someone who is participatory, inclusive, empathetic, non-patronising, caring and non-judgemental.
They should be open and honest, competent with expertise in the relevant areas. They should be credible and genuine and consistent in the message they give.

When communicating with the public and media it must be recognised that different populations have different information needs. Some of the methods that may be used include:

- Public meetings
- Speaking to community leaders and opinion formers
- Telephone help lines
- Personalised letters
- Developing newsletters and bulletins
- Using free papers or magazines

One of the most difficult risk communications can be where there is very little information, as is often the case at the beginning of an incident.
If we look at an airline situation where there has been a plane crash and there is little information to pass to relatives the spokesperson will always show concern and empathy, they will outline the steps being taken and the support they are giving. They will commit to providing further information and will back up their words with actions. If we contrast this with a public agency responding for the first time – commonly they offer a junior spokesperson who simply says that there is no information available. This presents a negative picture about the competence and commitment and demonstrates little in the way of empathy and openness.

Basic media requirements should always be considered, the media believe the public want information on what the situation means to them and the impact on their personal life and routine. They want access to more information via a helpline or other suitable means such as a website. They also want information on what is being done on their behalf to manage the situation.
One main media requirement is the press release, here it is important to clearly identify the problem at the outset if this is possible. It is important to answer the who, what, when, where, why and how questions in a strong leading paragraph. The first ten words are crucial in securing interest. Use a second paragraph to elaborate the content of the first and expand the details, summarise your key messages and give contact details for follow up. It

is also always helpful to supplement press releases with fact sheets/ FAQ's to provide supplementary information and anticipate the most likely follow on questions.

When dealing with interviews the following points are important for the person being interviewed:

- Keep calm and courteous, be collected and confident and consider the reporter's needs.
- Have a clear agenda and purpose, focus on getting your key message across
- See your role as helping the reporter in understanding the right message and passing this on to the public
- Make your point in 30 seconds and in no more than 90 words
- Don't make commitments or promises you can't keep and don't offer opinion beyond your competence
- Only discuss what you know not what you think
- Correct anything you get wrong quickly and openly
- Never lie to, embarrass or argue with a reporter, but also don't allow them to put words into your mouth.

6 SUMMARY

In summary the barriers to communicating effectively can be access to information, the risk communication process itself and general contextual issues.
It is very important therefore to know your audience, identify your key messages and stick to them, recognise that safety is relative and to build trust through mutual respect.

References

Health Protection Network. Communicating with the Public about Health Risks. September 2008
http://www.documents.hps.scot.nhs.uk/about-hps/hpn/risk-communication.pdf

POTABLE WATER CONTAMINATION EMERGENCY: THE ANALYTICAL CHALLENGE

B. May

The Food and Environment Research Agency, Sand Hutton, York, YO41 1LZ, UK

1 INTRODUCTION

In the event of a potable water contamination incident, a laboratory will be called upon to rapidly identify any contaminants present often with very little information as to the source of contamination.[1,2,3] With this in mind in 1997 the Laboratory Environmental Analysis Proficiency (LEAP) Emergency Scheme was established. The primary function of this scheme was to test the ability of a laboratory to analyse chemical samples for unknown contaminants from a simulated potable water contamination incident. The scheme is now administered by The Food and Environment Research Agency (FERA), which is an executive agency of UK Department for Environment, Food and Rural Affairs (DEFRA).

2 CONTAMINATION OF POTABLE WATER SUPPLY

Contamination of a potable water supply can occur by two main routes either accidentally or maliciously e.g. extortion or terrorism. Most contamination incidents in the UK have occurred accidentally; the 1988 Camelford water pollution incident is a prime example, and is to date perhaps Britain's worst water supply contamination incident.[4] However, the threat posed from malicious contamination cannot be ignored. Thankfully contamination incidents are rare events, however, past experiences have clearly shown that when they do occur they have two major implications for water companies. The first is financial and the second is that of public health. Whilst financial loss can be great, it is the public health concerns that tend to continue for many years after the event. Indeed, some twenty two years after the Camelford incident questions are still been raised on the potential long term health effects arising from this incident.

3 MAIN SCHEME OBJECTIVES

The main objectives of the scheme are to measure participating laboratories' ability to produce "fit for purpose" timely results during an emergency incident, with the long term aim to bring all participating laboratories up to the same high reporting standards. Clearly some laboratories will have more experience in the analysis of unknown contaminants than others.

4 SCHEME ORGANISATION

4.1 Sample Preparation

The exercise takes place over a 12 day period and begins with LEAP spiking a potable water sample with chemical contaminants. A combination of inorganic and organic contaminants is chosen, with a typical exercise containing four to seven contaminants. Sample spiking takes place on a Thursday, which is also the day of sample dispatch.

The following spiked samples are supplied:

- 2 x 1 litre in glass bottles for organics analysis
- 1 x 1 litre in a plastic bottle for inorganic analysis
- 1 x 250ml in a plastic bottle acidified to (0.1M HNO_3) for metals analysis

A set of blank samples is also supplied in order to allow the laboratory to compare the background matrix of the sample before any contaminants are added.

A separate radioactivity sample is also supplied and laboratories are requested to analyse this sample for gross alpha and beta radiation only.

4.2 Sample Dispatch

The samples are dispatched on Thursday for next day delivery to senior laboratory management who are requested to store all samples at $4^{o}C$ on receipt.

4.3 Contamination Scenario

Included with each set of samples is a contamination scenario, which details the circumstances of the contamination.

The contamination scenario asks the following questions: -

- Is there any significant contamination present
- If so, what is the approximate concentration (s) present
- What are the potential sources of contamination
- What analytical methods have been employed to detect the contaminant
- Were any screening tests used
- What can water be used for

4.4 Sample Analysis

In order for the participant to gain maximum benefit from participation in the exercise senior laboratory management is requested to ensure that the laboratory has no prior warning of delivery of samples or contamination scenario. Each participant is requested to present the samples and contamination scenario to the laboratory at 10.00 hours on Tuesday. Laboratories are also provided with a response sheet that requires the laboratory to record contaminants identified in test samples together with answers to other questions given in the contamination scenario. The laboratory is then requested to e-mail or fax their answers to LEAP as soon as any information becomes available.

4.5 LEAP Reporting

Participating laboratories are informed of contaminants present by e-mail shortly after the closing date for the exercise. This allows each laboratory to investigate its performance rapidly. A full confidential report is issued approximately one month from the exercise closing date.

4.6 User Forum Meeting

This meeting is held under Chatham House rules approximately one month after the closing date of the exercise. The meeting gives each laboratory an opportunity to discuss the outcomes of the exercise, any problems encountered during the analysis of the sample and share valuable analytical information on the best routes to detect contaminants. The minutes of these meetings are circulated to all exercise participants.

5 RESULTS AND DISCUSSION

5.1 Detection Rates

With time, the overall detection rates have generally improved as laboratories improve their screening techniques. Tables 1 to 3 show results obtained from three exercises in the period from March 2003 to December 2009.

Table 1 *Contaminants identified in exercise 6 (March 2003)*

Contaminants present	*Concentration ug/L*	*Correct identification*	
		number of labs (max = 15 labs)	*percentage (100% = 15 labs)*
Dimethoate	20	14	93%
2,4,5-T	20	6	40%
1,4-Dioxane	5,150	8	53%
Phenyl mercuric acetate	10	0	0%
Mercury	6	15	100%
Methanol	144,000	4	27%
Ethyl acetate	16,000	13	87%

Table 2 *Contaminants identified in exercise 11 (November 2005)*

Contaminants present	*Concentration ug/L*	*Correct identification*	
		number of labs (max = 20 labs)	*percentage (100% = 20 labs)*
Perchlorate (as K salt)	1000	0	0%
Rhenium	50	16	80%
Aldicarb	75	10	50%
Phosphamidon	70	20	100%
Glyphosate	202	1	5%
MCPA	10	11	55%
Aniline	339	20	100%

Table 3 *Contaminants identified in exercise 19 (December 2009)*

Contaminants present	*Concentration ug/L*	*Correct identification*	
		number of labs (max = 15 labs)	*percentage (100% = 15 labs)*
Mecoprop (MCPP)	100	14	93%
Imidacloprid	75	12	80%
2,4-D	50	14	93%
Glyphosate	350	8	53%
Fluoroacetate (as Na salt)	832	4	27%
Gallium	75	14	93%

5.2 Breakdown Products

Breakdown products form due to breakdown of the contaminant, which can be accelerated in the presence of chlorine in a potable water sample. Quite often it is these breakdown products, which are detected before the spiked contaminant is actually identified. And in some instances the breakdown products can be more toxic than the initial contaminant. Examples of some of the major breakdown products reported by participants are given in Table 4.

Table 4 *Breakdown products*

Contaminant spiked	*Breakdown products reported*
Phenol	Chlorophenols
	Bromophenols
Carbofuran	Carbofuran phenol
Aldicarb	Aldicarb nitrile
	Aldicarb sulfoxide
Mercaptoacetic acid	2-(Carboxymethyl disulfanyl) acetic acid
Dodecyldimethylammonium bromide	1-Chlorodecane
	Tertiary amines
2,4-D	2,4-Dichlorophenol

5.3 False Positives

In an emergency contamination incident the correct identification of a contaminant will be of paramount importance. The consequences arising from reporting a contaminant as present only to find out at a later date that it is actually another contaminant, the so called "false positive" result could be very costly. A number of spiked contaminants have resulted in laboratories reporting false positives. Table 5 gives some of the false positive results reported by participants. Figure 1 is a stacked plot, which shows a gas chromatography mass spectrometry (GC-MS) scan obtained by a participant for the spiked contaminant chloropicrin. Below is the National Institute of Standards and Technology (NIST) spectral library match obtained by the same participant for carbon tetrachloride which can be seen to be almost identical to that obtained for chloropicrin. This resulted in the laboratory initially reporting the contaminant present as carbon tetrachloride when it was actually chloropicrin. Subsequent headspace analysis for carbon tetrachloride showed that the sample did not contain any carbon tetrachloride.

Table 5 *False positives*

Contaminant spiked	*False positive reported*
Carbofuran	Methabenzthiazuron
1,4-Dioxane	Tetrahydro-3-fluoranol
Chloropicrin	Carbon tetrachloride

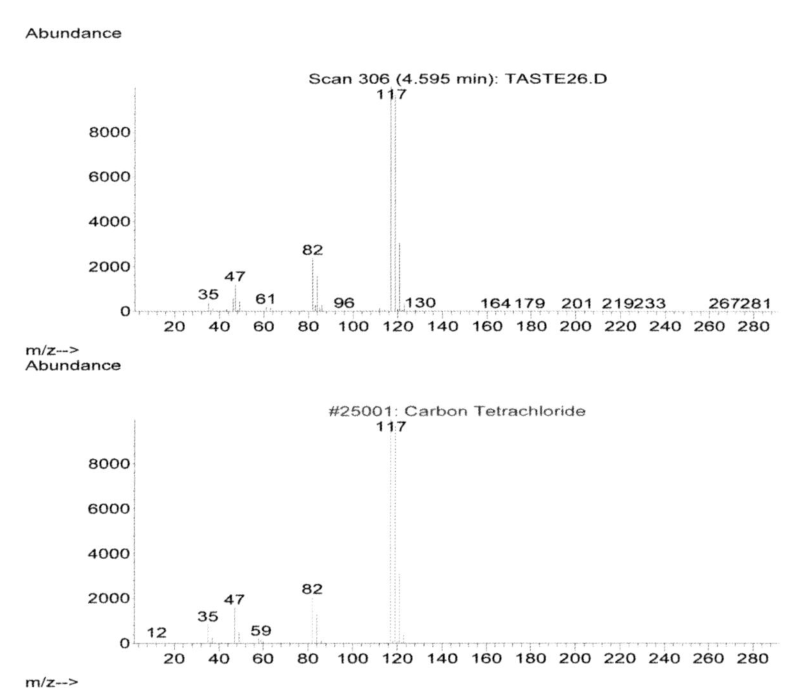

Figure 1 *Stacked plot showing above GC-MS scan for chloropicrin and below NIST spectral library match for carbon tetrachloride.*

5.4 Repeat Analysis

Some of the contaminants that were spiked were initially very poorly detected by all participants. In order to see if any lessons were learned from the first exercise these contaminants were again spiked in exercises carried out a number of years later. Table 6 gives a list of the contaminants that were spiked more than once over a number of years along with percentage (%) of laboratories that detected the contaminant. It can be seen that in most cases there is a significant improvement in the (%) of laboratories detecting the contaminants which had previously been poorly detected by all participants. Some contaminants still do seem to give poor overall detection rates, in particular sodium fluoroacetate where only a maximum of 27% of participating laboratories managed to detect this highly toxic contaminant. This substance is detectable using ion chromatography (IC) or liquid chromatography mass spectrometry time-of-flight (LC-MS- TOF) but can be easily missed as it only produces a very small peak using IC. Molecular ion peak is also difficult to detect amongst other peaks when using LC-MS - TOF.

Table 6 *Repeat analysis*

Contaminant	*Exercise*	*Date*	*Concentration ug/L*	*% of labs detecting contaminant*
Paraquat	7	Nov 2003	200	11
	14	May 2007	1000	65
	20	May 2010	400	67
Aldicarb	8	May 2004	50	28
	11	Nov 2005	75	50
Methanol	5	Nov 1999	16,000	0
	6	Mar 2003	144,000	27
	13	Nov 2006	44,200	76
	20	May 2010	79,000	60
1,4-Dioxane	6	Mar 2003	5150	53
	14	May 2007	3605	88
Glyphosate	11	Nov 2005	202	5
	19	Dec 2009	350	53
Chromium (VI)	7	Nov 2003	200	6
	12	May 2006	150	50
	17	Nov 2008	250	44
Fluoroacetate (as Na salt)	8	May 2004	2000	6
	14	May 2007	3000	24
	19	Dec 2009	832	27

6 CONCLUSION

The analytical challenge faced by the laboratory when presented with a potable water sample, where little or no information is available as to the source of the contamination is formidable. To get some perspective on the challenge, one only needs to look at the Chemical Abstracts Service (CAS) Registry that on the 7th September 2009 recorded its 50 millionth chemical substance.[5] These are just the chemical substances that have received a unique identifier, the CAS Registry Number (CAS RN).

Since its inception in 1997 LEAP has distributed twenty potable water simulated contamination exercises that have helped participating laboratories to produce clear strategies for dealing with emergency incident samples. The analytical results that have been obtained to date have shown a significant improvement in participating laboratories' abilities to detect unknown contaminants. Whilst detection rates have generally improved from the early days there is clearly still more room for improvement, which will hopefully come as lessons are continually learned from the simulated contamination exercises. The results reported have shown that inorganic determinands and metals are usually easily identified with organic non-polar substances of relatively low molecular weights being less well identified. Highly polar, involatile or thermally labile organic compounds can be poorly identified by participants. Contaminant concentrations are also often under or over estimated usually because results are reported from screening method only where no calibration standards were available for the contaminant at the time of analysis.

The reporting of breakdown products and false positive results have clearly emphasised the need to ensure further analysis is undertaken to confirm the presence of a contaminant. Breakdown products are especially likely to be formed in potable waters due to the presence of chlorine in the sample. False positive results like the carbon tetrachloride example given in this paper serve to show the importance of using where possible another analytical technique to back up the positive identification obtained from a routine screening method.

The key to improved rapid detection of organic contaminants lies with more efficient screening protocols for methods using both (GC-MS) and high resolution liquid chromatography and mass spectrometry (HR-LC-MS) or (LC/MS/MS). More recently nuclear magnetic resonance spectroscopy (NMR) has been used for both structural elucidation and rapid detection of organic contaminants in potable water.[6] Clearly there is no one analytical technique which will be able to detect all unknown contaminants in a potable water sample. It is the more efficient use of a number of different analysis techniques that will ultimately improve the ability of a laboratory to detect unknown contaminants. By participating in simulated contamination exercises such as the LEAP Emergency Scheme, laboratories will be able to test how well their emergency analytical procedures and protocols are working.

7 ACKNOWLEDGEMENTS

The Scheme would like to thank Prof K.C. Thompson, Chief Scientist at ALcontrol Laboratories, and G. Mills, Principal Development Scientist at Severn Trent Laboratories for their advice and support of the LEAP Emergency Scheme.

References

1 B.May, in *Water Contamination Emergencies Enhancing Our Response*, ed. K.C. Thompson and J. Gray, RSC, Cambridge, 1st edn, 2006, pp. 229 – 235.
2 K.C. Thompson and S. Scott, in *Water Contamination Emergencies Enhancing Our Response,* ed. K.C. Thompson and J. Gray, RSC, Cambridge, 1st edn, 2006, pp. 216 – 228.
3 G. O Neill, C. Ridsdale, K. C. Thompson and K. Wadhia, in *Water Contamination Emergencies: Can we cope?*, ed. K. C. Thompson and J. Gray, RSC, Cambridge, 1st edn, 2004, pp. 100-109.
4 Consultation Report by *Committee on Toxicity of Chemicals in Food, Consumer Products and the Environment Subgroup Report on the Lowermoor Water Pollution Incident,* January 2005. *cot.food.gov.uk/pdfs/lowermoorreport05.pdf.*
5 M. Toussant, *Chemistry World,* A Scientific Milestone, guest editorial by Matthew Toussant, 2009 December, **6**, No 12.
6 A.J. Charlton, J.A. Donarski, S.A. Jones, B.D. May, K.C.Thompson, *Journal of Environmental Monitoring*, 2006, **8**, 1106.

SOME EXAMPLES OF THE OPERATION AND BENEFITS OF THE UK WATER LABORATORIES' MUTUAL AID SCHEME

K. Clive Thompson[1] P. Frewin[2] and T. Brooks[2]

[1]Chief Scientist, ALcontrol Laboratories, Rotherham, South Yorkshire, S60 1FB, United Kingdom E-mail clive.thompson@alcontrol.com
[2]Senior Radiochemists, South West Water, Exeter Laboratory, Exeter, EX2 7AA

1 INTRODUCTION

The UK water laboratories mutual aid scheme (WLMAS) was originally set up in 1995 after all the major UK water laboratories met in Rotherham. It has continuously evolved and improved its capabilities since that time. It is an informal group that was set up for laboratories involved in the analysis of emergency incidents primarily relating to drinking waters and rivers. Although water companies had set up groups to deal with emergency incidents there was not any group that dealt with the analysis aspect with respect to UK laboratories. The group's main objectives are: -

- To continuously improve the response capability of laboratories carrying out emergency incident analysis
- To share information
- To initiate mutual new aid developments / initiatives
- To share positive and negative experiences.
- To identify and adopt best practices
- To provide reassurances to the water industry that the laboratory services can cope with emergencies relating to chemical and radioactivity pollution incidents.
- To provide contingency event cover in the event of a major catastrophe occurring on one of the group member laboratories

The introduction of the WLMAS fifteen years ago has greatly enhanced the UK analytical capabilities to efficiently handle emergency potable water incidents. This informal self-help group meets on an annual basis at a different member location each time. There are no charges or fees for membership and each participant laboratory has to bear its own costs. At the annual meeting, there are three or four short presentations on new techniques; improvement to existing techniques and any other areas considered to be of interest. There are no legal agreements or company lawyers involved. All meetings are held under Chatham House rules which state that when a meeting, or part thereof, is held under the Chatham House Rule, participants are free to use the information received, but neither the identity nor the affiliation of the speaker(s), nor that of any other participant, may be revealed. Sometimes members are happy to allow this information to be given.

Three specialist associated sub-groups have also been set up: -
Organic analysis; Laboratory Environmental Analysis Proficiency (LEAP) Emergency Scheme (two exercises per year on random dates to help ensure emergency analysis staff training is kept up to date) and Rapid radioactivity screening. The latter radioactivity screening group has developed a rapid gross alpha/gross beta method and comprehensively validated it. This method will allow a laboratory to be able to screen over 150 drinking water samples per day in an emergency situation

2 UK WATER LABORATORIES MUTUAL AID SCHEME (WLMAS)

2.1 UK Water Laboratories Mutual Aid Scheme (WLMAS) Remit

The remit includes the following items; -

- Discussion of all analytical issues relating to handling emergency incidents by annual meeting and E-mail;
- To cover mainly potable water, but also non-potable water incidents;
- Improvement of liaison between laboratories and emergency planning sections of water companies;
- Maintaining a regularly updated network of contacts;
- Maintaining an up to date 24h/365d emergency response capability statements for all relevant labs;
- Standardising the response to emergency incidents or sudden/planned loss of laboratory facilities;
- Ensuring that all mutual aid is on a reasonable endeavours basis and only charged at normal commercial rates;
- To continuously improve the response capability of laboratories carrying out emergency incident analysis;
- To share information;
- To initiate mutual new aid developments / initiatives ;
- To share both positive and negative experiences;
- To identify and adopt best practices;
- To provide reassurances to the water industry that the laboratory services can cope with emergencies including CBRN;
- To provide contingency event cover in the event of a major catastrophe occurring on one of the group member laboratories;

2.2 WLMAS Specific Objectives

- Screen 100 samples in 2 hours for relevant chemical parameters
- Screen 10 samples in the first 3 hours followed by 10 samples every 30 minutes for gross alpha and beta radioactivity (assuming sufficient trained staff resources)
- Maintain ‘experts’ network
- Continuously review and update current capability and identify best practice
- Participate and learn from the Fera LEAP emergency proficiency scheme[1,2]
- Develop relevant robust rapid screening tests
- Identify and investigate future developments and new techniques
- Include more potential target compounds on suites

- Consider utilisation of shared resources between companies (e.g. Cryoprobe NMR)
- Set up a secure WLMAS website

2.3 Key Issues

Emergency response planning for fit for purpose analysis of samples associated with very high impact very low probability events is notoriously difficult. This is particularly true when dealing with the laboratory analysis arising from potable water emergency pollution incidents. It is very difficult to convincingly prove a negative (no contamination) quickly in an emergency situation given the relatively large number of potential toxic chemical substances.

To date, serious industrial, agricultural and natural emergency pollution incidents are the most common cause of emergency incidents rather than malicious contamination. There is no silver bullet yet for comprehensive rapid screening of emergency incident water samples, despite large amounts of money being spent by the various national bodies. In June 2010 the US approved a $44 billion fiscal 2011 spending bill for the Homeland Security Department.

A number of commercial rapid test kits (some aimed at domestic consumer use) are available with many exhibiting a significant percentage of false negative and positive results rendering them effectively useless in life threatening emergency situations. (See below)

Most routine water laboratories are geared to high volume targeted analysis (in the UK some water labs are handling 500+ drinking water samples per day.) Targeted analysis does not require in depth knowledge of analysis by the analyst. However, "absence (screening) analysis" for a wide range of diverse parameters is far more complicated than "targeted analysis" and requires skilled and experienced analysts. This is particularly true for ultra trace organic and radiological emergency analysis.

The key laboratory emergency testing issues are considered to be: -

- *Adequate resource allocation (both trained staff & necessary equipment)*
  The vast majority of the emergency analysis is carried out using routine equipment. It is very difficult in the current financial climate to justify dedicated leading edge equipment to handle very low frequency major emergency incidents. Thus when incidents do occur, this can result in major disruption to the routine analysis work.

- *Maintaining capable fit for purpose 24 hours/365 days cover on a long-term basis*
  Maintaining a highly skilled adequate pool of staff to be able to effectively deliver 24 hours/365 days cover via a rota system is becoming increasingly difficult in the current financial climate. Some of the available emergency analysis procedures may only be used in anger only once or twice a year. Maintaining staff training under these conditions can be demanding. The LEAP emergency proficiency scheme[1,2] is a useful way of assessing staff capability for competently dealing with emergency samples. Also when a major prolonged incident does occur, it is very difficult to maintain the routine analysis and the emergency analysis workloads with the available staff resources. However, some of the routine work can be sub-contracted to other members the scheme. All members are accredited to carry out most of this work.

- *Effective handling of the time-pressure for results issue.*
  Frequent urgent requests from operational staff for emergency analysis results can rapidly adversely effect analytical staff performance and result in a very stressful environment. Laboratories should clearly indicate to operational staff when explicit urgent results are expected to be available to avoid analyst stress and potential associated analysis errors. The key initial question is can any significant change in the water be detected even if the nature of the substance(s) responsible for the change cannot immediately be identified?

- *Number of relevant substances*
  Although there over 30 million known organic substances, there are probably less than 2000 relevant organic and less than 100 relevant inorganic significant risk agent substances. Efforts have been made to try and ensure that the screening methods used by members of the group will cover the major risk agents as perceived by the group. A problem with chemical testing is that there is a very large number of potentially toxic substances and it is very difficult to prove a negative particularly under pressure in an emergency situation involving large numbers of samples. A complementary approach, whilst running chemical screening, is to employ simple rapid screening Ecotoxicity toxicity testing to detect acute toxicity. This is a holistic approach that detects toxic effects on organisms employed as biological indicators in the test used.[3]

- *Potential wide range of radioactive isotopes*
  The rapid gross alpha/beta method developed by the group (see subsequent chapter) should be able to detect excessive levels of all potential risk agent isotopes except 75Se, 95Nb, 103Ru or 169Yb. Some of these radionuclides do not emit beta particles, while in the other cases the energy of the beta particle emission is too low to be detected by the method used.[4]

- *What does "Nothing significant found" signify? Is this analysis response consistent across all labs?*
  This is the key question and the results from the LEAP Emergency proficiency scheme would indicate that this not all labs detect all potential substances of interest. This proficiency scheme[1,2] distributes a sample containing up to seven different risk agents to all participating laboratories at a random date (unknown to the laboratories) and assessment is based upon rapid detection of the added substances with a semi-quantitative estimate of their concentrations. There are two rounds per annum. The results from this scheme generally indicate a gradual improvement of the participating laboratories which includes all WLMAS group members. There are wash up meetings to discuss the results of each exercise and the screening procedures used by the various laboratories. This allows laboratories which missed some substances to consider alternative improved detection techniques.

  The WLMAS group has recommended that periodic analysis of uncontaminated final treated water samples from all major supplies should be undertaken in order to determine the natural concentration levels of natural (harmless) matrix constituents. This particularly applies to any natural (harmless) organic substances (e.g. related to humic/fulvic acids) that are detected (but are normally not identified) when using the LC and GC chromatographic emergency methods. Thus if any of these harmless substances are detected in an emergency sample at the levels found in uncontaminated samples, then these substances can be safely ignored.

- *Importance of avoiding false negatives and minimising false positives when using rapid screening methods*
  Rizak and Hrudey[5] have shown that if the hazard that monitoring evidence is screening for is rare (as one would expect in monitoring treated water quality), the chance of any positive detected being a true positive, the positive predictive value (PPV) will be small unless the false positive detection rate is as small or lower than the frequency of that hazard.
  This quantitative reality depends on the limited capabilities of methods and the frequency of the hazards. This reality must be recognized so that monitoring programs can be planned effectively. Responses to positive monitoring results must be appropriate to their likely meaning to ensure that responses will do more good than harm. The above authors then cite an example of a hypothetical test that one hazard screening assay manufacturer claimed a false-positive rate (a) of only 3% for the pathogen *Bacillus anthracis*. Intuitively, the 3% false positive rate appears good for a rapid screening test. However, suppose that one were to apply such screening technology to circumstances of rare contamination where, for the sake of illustration, only 1 out of 300 samples are truly hazardous. One can ask: If we get a positive result (contaminant is detectable or exceeds the standard) from an analytical test, how likely is that "positive" result to be correct?[6,7] One will need to screen 299 samples free of detectable levels of the contaminant to find the one sample that contains detectable levels. With a false positive rate of 3%, approximately nine false positives will be found whilst searching for the one true positive. Consequently, only 10% of positive results from this analytical test will correctly reflect the presence of a hazard (i.e., the positive predictive value (PPV) will be ~10%). This finding is an inescapable reality (as a function of the false positive rate and frequency of true hazards) for any analytical screening test. As the hazard becomes more rare, one can expect false positives to exceed true positives unless a test offers a false positive rate somewhat less than the frequency of the hazard[3]

3 EMERGENCY RESPONSE CAPABILITY STATEMENTS

It was recognised at the very start of the WLMAS that it was essential to standardise the response to a major incident which may even involve the loss of the company's laboratory facilities. In order to achieve this it was agreed to set up and maintain, an up to date database of Emergency Response Capability Statements. These statements list concisely, the contact details and the analytical capabilities of the laboratory. This would allow other members to identify supportive facilities if and when needed. The mutual aid would be provided on a reasonable endeavours basis and would not be contractual.

All charges would only be at normal commercial rates. Several laboratories are commercial competitors therefore a ground rule that was established early on was the agreement to operate under Chatham House rules (see above). Figure 1 shows the information required on the capability statements. These are regularly updated on a formal basis by a volunteer member of the group. All emergency work undertaken would be on a reasonable endeavours basis. Some adjacent area water companies have set up localised

Emergency Response Capability Statement Pro Forma

Company Name: [ ] Date form produced/updated [ ]

Laboratory Name: [ ]

Address: [ ]

CLICK HERE TO RETURN TO MASTER SHEET

Post Code: [ ]

Prime Contact [ ] E-mail [ ]

Other Contacts [ ] E-mail [ ]

Tel No. Working hours: [ ] FAX No.: [ ]

Tel No. Outside working hours: [ ] (24 hours, 365 days Y/N) [ ]

The laboratory carries out the routine analysis of the parameters in the Drinking Water Directive to the standard required by DWI Y/N [ ]

The laboratory is UKAS accredited Y/N [ ]

Specialist Analysis Areas:-

Radioactivity (Gross alpha and beta): Y/N [ ]
Radioactivity gamma ray spectrometry* Y/N [ ] * If yes, please complete form "EMERFORCii.2001" for specialist radioactivity
Radioactivity other (specify):

[ ]

Specialist microbiology Y/N [ ]
Regulatory Cryptosporidium Y/N [ ] Viability Y/N [ ]
Giardia Y/N [ ]
PCR capability Y/N [ ]
Flow cytometry Y/N [ ]
Class 3 pathogens Y/N [ ]
Virology Y/N [ ]
Outline of specialist microbiology capabilities:

[ ]

Typical capacity for extracting and running unknown methylene dichloride extracts for identification/semi-quantification by GC-MS in incident situations Samples/day [ ]

Low resolution (GC-MS) Y/N [ ]
High resolution (GT 5000) GC-MS Y/N [ ]
LC-MS Y/N [ ]
LC-DAD Y/N [ ]
Purge and trap GC-MS Y/N [ ]

Odour identification Y/N [ ]
Gas identification Y/N [ ]
Air monitoring sampling Y/N [ ]
Air monitoring analysis Y/N [ ]
Ecotoxicity testing Y/N [ ]

Outline of specialist organics and ecotoxicity testing services and any other comments on the above:

[ ]

Human / animal / invertebrate toxicity information databases available Y/N [ ]

Outline of toxicity information services available:

[ ]

Other services Y/N [ ]
Outline of other services offered:

[ ]

Emergency incident contact procedure outside working hours:

[ ]

Please note:
(1) Emergency incident response is on a reasonable endeavours basis.

If the Core Business of a water laboratory offering this service has an emergency situation, it must take priority.

(2) All commercial costs should be agreed between parties prior to sending samples.

(3) Any changes to the above information, please contact:-

Figure 1 *Emergency response capability statement proforma form*

sub-groups with a more detailed capability statement that also specifies the maximum workload of all relevant parameters that could be undertaken at short notice. This would involve ceasing some less essential routine work for their own water company. This reduction has been agreed with the relevant water company. This would also cover a contingency event where a laboratory suffered a catastrophe and either ceased to operate (e.g. a major fire) or had key sections shut down (e.g. a localised fire)

4 DEVELOPMENT OF A RAPID GROSS ALPHA AND BETA SCREENING METHOD FOR THE WATER INDUSTRY

The 2000 European Directive requires that a Total Indicative Dose be assessed (excluding Tritium, K-40, Radon and its decay products). If gross alpha is <0.1 Bq/l and gross beta <1 Bq/l, no further analysis is required and the water sample can be considered to be wholesome with respect to radiological parameters (excluding Tritium, Radon and its decay products.)

In the event of nuclear accident, terrorist attack or other similar event Water Companies will need to rapidly ascertain the health and safety risk relating to radioactivity. The development of appropriate capability is described in a subsequent chapter in this book

5 CONCLUSIONS

- The setting up of the WLMAS in 1995 and its three sub-groups on organic analysis; radioactivity analysis and the Fera Emergency proficiency scheme has significantly improved emergency response analysis in the UK.
- The introduction of the regularly updated 24h/365d emergency response capability statements for all relevant labs has proved very successful. The implementation of a reasonable endeavours basis, absence of any formal legal agreement and associated lawyers has been found to work well.
- Significant progress has been and continues to be made; however, there are still limitations to what a water laboratory can achieve. It is important to appreciate that the laboratory will not always be capable of detecting every contaminant. However significant changes in surrogate parameters such as electrical conductivity; TOC; colour, odour, uv scan, ammonia, chlorine demand, pH etc. can often indicate the presence of other undesirable substances. The efficient interpretation of the complete sample analysis results is an area where expert knowledge is essential.
- Detecting one contaminant does not preclude the presence of others. Rapid analysis is still not instantaneous – operational staff need to be aware of a credible timeline and kept regularly updated with progress and expected time of results availability. It is essential for the Operational Staff to prioritise samples for analysis to ensure that key samples are analysed before less important samples.
- It is also important to understand that the identification of unknowns is not an exact science. There should be no expectation that any combination of technology or analytical capability will always guarantee 100% success. However, the vast majority of samples will usually prove not to be contaminated.

- A useful complementary approach, whilst running chemical screening, is to simultaneously employ simple rapid screening ecotoxicity toxicity testing to detect acute toxicity.
- What is needed for the future are rapid, automated analytical techniques that are capable of detecting very wide range of unknown organic substances especially non-polar substances. HPLC;MS/MS and HPLC-TOF are showing greatest promise in this area. The cost of the instrumentation has fallen significantly and the ease of use has improved dramatically.
- The use of NMR techniques and especially dynamic nuclear polarisation (DNP) NMR[8,9] would appear to have considerable potential as a highly specialised back up method that would be employed in a single national laboratory. This technique has the potential to detect a very wide range of organic substances down to ppb levels and even if the detected substance is not in the system library, some idea of the structure can often be deduced. The NMR technique should be regarded as a back up technique for chromatographic methods. (It is especially useful for substances that do not chromatograph well and/or are difficult to ionise.)

References

1 Barry May, *Laboratory environmental analysis proficiency (leap) emergency scheme in, Water Contamination Emergencies Enhancing Our Response,* ed. K.C. Thompson and J. Gray, RSC, Cambridge, 2006, pp. .229-235.

2 Barry May, *Potable water contamination emergency: -the analytical challenge, in Water Contamination Emergencies IV, monitoring, understanding, acting* RSC, Cambridge, 2011, pp 110–116.

3 K. C. Thompson, *Rapid methods, in Water Contamination Collective Responsibility* ed. K.C. Thompson and J. Gray, RSC, Cambridge, 2008, pp 267-292.

4 Health Protection Agency, 2008. *Handbook for Assessing the Impact of a Radiological Incident on Levels of Radioactivity in Drinking Water and Risks to Operatives at Water Treatment Works.* [Online] (Updated 1 September 2008) Available at: http://www.hpa.org.uk/Publications/Radiation/HPARPDSeriesReports/ HPARPD040/[Accessed 3 November 2010].

5 S. Rizak and S. E. Hrudey, *Improved understanding of water quality monitoring: evidence for risk management decision-making, in Water Contamination Emergencies Enhancing Our Response*, ed. K.C. Thompson and J. Gray, RSC, Cambridge, 2006, pp. 350-354.

6 S. E. Hrudey and S. Rizak. *J. Am. Water Works Assoc*., 2004, **96**, 110.

7 S. Rizak and S. E. Hrudey, in *Proceedings of the 11th Canadian National Conference and 2nd Policy Forum on Drinking Water*, Calgary, Alberta, Canada, April 3–6, 2004.

8 A. J. Charlton, J A. Donarski, B D. May and K. C. Thompson, *Optimisation of NMR methodology for non-targeted detection of water contaminants, in Water Contamination Emergencies Collective Responsibility,* ed. K.C. Thompson and J. Gray, RSC, Cambridge, 2008, pp 245-251.

9 A.J. Charlton, J.A. Donarski, S.A. Jones, B.D. May, K.C. Thompson, *Journal of Environmental Monitoring*, 2006, **8**, 1106.

DEVELOPMENT OF A RAPID GROSS ALPHA & BETA METHOD FOR THE WATER INDUSTRY

T. Brooks[1], P. Frewin[2] and K. C. Thompson[3]

[1]Senior Radiochemist, South West Water, Exeter Laboratory, Exeter, EX2 7AA
[2]Consultant Radiochemist, phillipafrewin@aol.com
[3]Chief Scientist, ALcontrol Laboratories, Rotherham, South Yorkshire,
S60 1FB, United Kingdom E-mail clive.thompson@alcontrol.com

1 INTRODUCTION

In April 1986 the world awoke to the news of the Chernobyl disaster. The resultant radioactive cloud affected much of Europe and this increased the need for environmental monitoring for radioactivity. Laboratories were swamped by the vast number of samples and food, livestock, air and land monitoring took priority over water. In some cases results for water samples took six months to come back.

The UK water industry recognised that in-house monitoring was essential for routine and emergency situations. This chapter will look at the development of a rapid gross alpha and beta method for emergency use.

By 2000, following European legislation[1], UK water quality regulations[2] included a requirement for routine radioactivity to monitor total indicative dose and tritium for the first time. Prior to this it was covered by the World Health Organisation (WHO) requirements for wholesomeness in drinking water[3].

2 BACKGROUND

The Drinking Water Directive (DWD), Council Directive 98/83/EC[1] set very stringent limits for gross alpha and beta in drinking water before further isotopic quantification is required. At this time only three UK water laboratories had the capability to carry out gross alpha and beta analysis, but they needed to contract out samples for isotopic speciation if required.

This is appropriate for drinking water over a lifetime, but not so during a radioactive incident. The National Radiological Protection Board (HPA) therefore recommended Action Levels[4] for potential use following the deliberate or accidental use of radioactivity, for short term water consumption, based on Council of Europe intervention levels for liquid food (CFILs)[5].

An NRPB (HPA) study has recommended Action Levels for UK drinking water following emergency incidents. These levels are based on Council Food Intervention Levels (CFILs) for liquid food published by the Council of European Communities. The proposed "acute" screening levels are 5 Bq/l for gross alpha and 30 Bq/l for gross beta. A subsequent Health Protection Agency (HPA) report[6] concluded that samples measuring below the screening level would not exceed these Action Levels.

From these Action Levels for individual isotopes, gross alpha and beta screening levels[7] for use during an emergency were proposed, at 5 Bq/l for gross alpha and 30 Bq/l for gross beta. These levels were evaluated using the British Standard methods[8], and it was concluded that in nearly all cases measurements below screening levels would not exceed the Action Levels[9].

Table 1 *Recommended UK Action Levels for drinking water supplies*[a] [6]

Radionuclide	Action levels[b] ($Bq\ l^{-1}$)	Categorisation of radionuclides considered in Handbook[c,d]
Isotopes of strontium, notably ^{90}Sr	125	^{90}Sr
Isotopes of iodine, notably ^{131}I	500	^{131}I
Alpha-emitting isotopes of plutonium and transplutonium elements	20	^{238}Pu, ^{239}Pu, ^{241}Am
All other radionuclides of half-life greater than 10 days, notably radioisotopes of	1000	^{60}Co, ^{75}Se, ^{95}Zr ^{95}Nb, ^{99}Mo, ^{103}Ru, ^{106}Ru, ^{132}Te, ^{134}Cs ^{136}Cs, ^{137}Cs, ^{140}Ba,
Caesium and ruthenium[e]		^{140}La, ^{144}Ce, ^{169}Yb ^{192}Ir, ^{226}Ra[f]

Notes:

a) These Action Levels refer to all water supplies which are intended, at least in part, for drinking and food preparation purposes.

b) It is the sum of the concentrations of all the radionuclides included within a category and detected in the water which should be compared with the Action Level.

c) The radionuclides considered are listed in Table 7 in reference 6.

d) For ^{235}U, action would be taken on the chemical toxicity of uranium which is of more concern to health than the radioactive content of the water [WHO, 2003].

e) This category does not include ^{14}C, ^{3}H or ^{40}K.

f) It should be noted that radon is unlikely to be a problem because it is very unlikely the deliberate contamination of a water supply with ^{226}Ra will lead to radon gas being produced.

Table 2 *Committed effective doses the consumption of tap water[a] contaminated at the Action Levels[b] [6]*

Radio-nuclide	Half-life[c]	Committed effective dose, mSv, following consumption for:								
		1 week			1 month			1 year[d]		
		1 yr old	10 yr old	Adult	1 yr old	10 yr old	Adult	1 yr old	10 yr old	Adult
^{60}Co	Long	$9\ 10^{-2}$	$4\ 10^{-2}$	$3\ 10^{-2}$	$4\ 10^{-1}$	$2\ 10^{-1}$	$1\ 10^{-1}$	5	2	1
^{75}Se	Long	$4\ 10^{-2}$	$2\ 10^{-2}$	$2\ 10^{-2}$	$2\ 10^{-1}$	$1\ 10^{-1}$	$8\ 10^{-2}$	2	1	1
^{90}Sr	Long	$3\ 10^{-2}$	$3\ 10^{-2}$	$3\ 10^{-2}$	$1\ 10^{-1}$	$1\ 10^{-1}$	$1\ 10^{-1}$	2	2	1
^{95}Zr	Long	$2\ 10^{-2}$	$7\ 10^{-3}$	$7\ 10^{-3}$	$8\ 10^{-2}$	$3\ 10^{-2}$	$3\ 10^{-2}$	1	$4\ 10^{-1}$	$4\ 10^{-1}$
^{95}Nb	Long	$1\ 10^{-2}$	$4\ 10^{-3}$	$4\ 10^{-3}$	$5\ 10^{-2}$	$2\ 10^{-2}$	$2\ 10^{-2}$	$4\ 10^{-1}$	$2\ 10^{-1}$	$2\ 10^{-1}$
^{99}Mo	Short	$1\ 10^{-2}$	$4\ 10^{-3}$	$5\ 10^{-3}$	$5\ 10^{-2}$	$2\ 10^{-2}$	$2\ 10^{-2}$	$4\ 10^{-2}$	$1\ 10^{-2}$	$1\ 10^{-2}$
^{103}Ru	Long	$2\ 10^{-2}$	$6\ 10^{-3}$	$6\ 10^{-3}$	$7\ 10^{-2}$	$2\ 10^{-2}$	$2\ 10^{-2}$	$7\ 10^{-1}$	$3\ 10^{-1}$	$3\ 10^{-1}$
^{106}Ru	Long	$2\ 10^{-1}$	$6\ 10^{-2}$	$5\ 10^{-2}$	$7\ 10^{-1}$	$2\ 10^{-1}$	$2\ 10^{-1}$	8	3	3
^{131}I	Short	$3\ 10^{-1}$	$1\ 10^{-1}$	$8\ 10^{-2}$	1	$4\ 10^{-1}$	$4\ 10^{-1}$	3	$9\ 10^{-1}$	$8\ 10^{-1}$
^{132}Te	Short	$1\ 10^{-1}$	$3\ 10^{-2}$	$3\ 10^{-2}$	$4\ 10^{-1}$	$1\ 10^{-1}$	$1\ 10^{-1}$	$4\ 10^{-1}$	$1\ 10^{-1}$	$1\ 10^{-1}$
^{134}Cs	Long	$5\ 10^{-2}$	$5\ 10^{-2}$	$1\ 10^{-1}$	$2\ 10^{-1}$	$2\ 10^{-1}$	$6\ 10^{-1}$	3	3	7
^{136}Cs	Short	$3\ 10^{-2}$	$2\ 10^{-2}$	$2\ 10^{-2}$	$1\ 10^{-1}$	$7\ 10^{-2}$	$1\ 10^{-1}$	$5\ 10^{-1}$	$3\ 10^{-1}$	$3\ 10^{-1}$
^{137}Cs	Long	$4\ 10^{-2}$	$4\ 10^{-2}$	$1\ 10^{-1}$	$2\ 10^{-1}$	$2\ 10^{-1}$	$4\ 10^{-1}$	2	2	5
^{140}Ba	Short	$6\ 10^{-2}$	$2\ 10^{-2}$	$2\ 10^{-2}$	$3\ 10^{-1}$	$9\ 10^{-2}$	$8\ 10^{-2}$	$9\ 10^{-1}$	$3\ 10^{-1}$	$3\ 10^{-1}$
^{140}La	Short	$4\ 10^{-2}$	$2\ 10^{-2}$	$2\ 10^{-2}$	$2\ 10^{-1}$	$7\ 10^{-2}$	$6\ 10^{-2}$	$8\ 10^{-2}$	$3\ 10^{-2}$	$3\ 10^{-2}$
^{144}Ce	Long	$1\ 10^{-1}$	$4\ 10^{-2}$	$4\ 10^{-2}$	$6\ 10^{-1}$	$2\ 10^{-1}$	$2\ 10^{-1}$	7	2	2
^{169}Yb	Short	$2\ 10^{-2}$	$6\ 10^{-3}$	$5\ 10^{-3}$	$7\ 10^{-2}$	$2\ 10^{-2}$	$2\ 10^{-2}$	$6\ 10^{-1}$	$2\ 10^{-1}$	$2\ 10^{-1}$
^{192}Ir	Long	$3\ 10^{-2}$	$1\ 10^{-2}$	$1\ 10^{-2}$	$1\ 10^{-1}$	$5\ 10^{-2}$	$5\ 10^{-2}$	2	$6\ 10^{-1}$	$6\ 10^{-1}$
^{226}Ra	Long	3	3	2	$1\ 10^{-1}$	$1\ 10^{-1}$	9	$2\ 10^{-2}$	$2\ 10^{-2}$	$1\ 10^{-2}$
^{235}U	Long	$4\ 10^{-1}$	$3\ 10^{-1}$	$4\ 10^{-1}$	2	1	2	$2\ 10^{-1}$	$1\ 10^{-1}$	$2\ 10^{-1}$
^{238}Pu	Long	$3\ 10^{-2}$	$2\ 10^{-2}$	$3\ 10^{-2}$	$1\ 10^{-1}$	$8\ 10^{-2}$	$2\ 10^{-1}$	1	1	2
^{239}Pu	Long	$3\ 10^{-2}$	$2\ 10^{-2}$	$4\ 10^{-2}$	$1\ 10^{-1}$	$9\ 10^{-2}$	$2\ 10^{-1}$	1	1	2
^{241}Am	Long	$2\ 10^{-2}$	$2\ 10^{-2}$	$3\ 10^{-2}$	$1\ 10^{-1}$	$7\ 10^{-2}$	$1\ 10^{-1}$	1	$9\ 10^{-1}$	2

Notes:

a) Consumption rates for tap water: 1 year old = 172 l y^{-1}, 10 year old = 197 l y^{-1}, Adult = 391 l y^{-1} [NRPB, 1994]. If site specific data on tap water consumption rates are available, values in the Table can be scaled directly to reflect different consumption rates.

b) See Table 3 in reference 6 for Action Levels.

c) Half-life: short = < 3 weeks; long: = > 3 weeks.

d) For short-lived radionuclides (half-life <1 month) the committed effective dose after 1 year of ingestion was calculated for a period equivalent to 8 radioactive half-lives.

Table 3 *Committed effective doses from 1 year's consumption of Drinking water initially contaminated at 1Bq l^{-1a} 6*

Radionuclide	Committed effective dose, mSv		
	1 year old	10 year old	Adult
^{60}Co	4 10^{-3}	2 10^{-3}	1 10^{-3}
^{75}Se	9 10^{-4}	5 10^{-4}	4 10^{-4}
^{90}Sr	1 10^{-2}	1 10^{-2}	1 10^{-2}
^{95}Zr	2 10^{-4}	9 10^{-5}	9 10^{-5}
^{95}Nb	8 10^{-5}	3 10^{-5}	3 10^{-5}
^{99}Mo	7 10^{-6}	2 10^{-6}	3 10^{-6}
^{103}Ru	1 10^{-4}	5 10^{-5}	4 10^{-5}
^{106}Ru	6 10^{-3}	2 10^{-3}	2 10^{-3}
^{131}I	1 10^{-3}	3 10^{-4}	3 10^{-4}
^{132}Te	7 10^{-5}	2 10^{-5}	2 10^{-5}
^{134}Cs	2 10^{-3}	2 10^{-3}	6 10^{-3}
^{136}Cs	8 10^{-5}	4 10^{-5}	6 10^{-5}
^{137}Cs	2 10^{-3}	2 10^{-3}	5 10^{-3}
^{140}Ba	2 10^{-4}	6 10^{-5}	5 10^{-5}
^{140}La	1 10^{-5}	5 10^{-6}	5 10^{-6}
^{144}Ce	4 10^{-3}	1 10^{-3}	1 10^{-3}
^{169}Yb	1 10^{-4}	4 10^{-5}	4 10^{-5}
^{192}Ir	4 10^{-4}	2 10^{-4}	2 10^{-4}
^{226}Ra	2 10^{-1}	2 10^{-1}	1 10^{-1}
^{235}U	2 10^{-2}	1 10^{-2}	2 10^{-2}
^{238}Pu	7 10^{-2}	5 10^{-2}	9 10^{-2}
^{239}Pu	7 10^{-2}	5 10^{-2}	1 10^{-1}
^{241}Am	6 10^{-2}	4 10^{-2}	8 10^{-2}

Note:

a) Consumption rates for tap water: 1 year old = 172 l y^{-1} ,10 year old = 197 l y^{-1}, Adult = 391 l y^{-1} [NRPB, 1994]. If site specific data on tap water consumption rates are available, values in the Table can be scaled directly to reflect different consumption rates.

3 RAPID ANALYSIS PROJECT

The Rapid Analysis Project was initiated in 2002 through the Mutual Aid scheme, and set the participating water companies the task of developing rapid screening methods for use during emergencies.

South West Water and Thames Water were set the objective of developing a rapid radioactivity method to meet the following goals:

1) To use existing water company analytical capability.
2) To analyse 10 samples within 3 hours of arrival at the laboratory.
3) To determine whether the water is fit for consumption.

In an emergency situation the routine gross alpha and beta method has limited capability as the turnaround of sample is very slow with instrumental analysis time the rate determining step.

4 DEVELOPMENT

The procedure for the routine gross alpha and beta method is as follows[8,10]

1) One litre of acidified sample is evaporated to dryness.
2) The residue is then sulphated, transferred to basin and bulking agent added.
3) It is then ignited at 350°C for one hour.
4) The sulphated residue is then ground and planchetted.
5) The planchette is then counted on a gas proportional counter for a minimum of eight hours.

It is impossible to turnaround samples on the day of receipt; and in the case of a large batch of samples it may take several days.

Development of the rapid method involved both laboratories evaluating where efficiencies could be made in their routine methods to achieve the time savings required. The two laboratories compared their existing methods, identifying the component stages. These stages were then individually scrutinised for potential short cuts.

The areas targeted were: reduced sample volumes (50ml, 100ml, & 200ml), different ashing times, and shorter counting times (30, 60 and 90min). All potential combinations were analysed to determine the optimum rapid method, which was then validated by spiking hard and soft waters, with the staff and existing instrumentation of both laboratories.

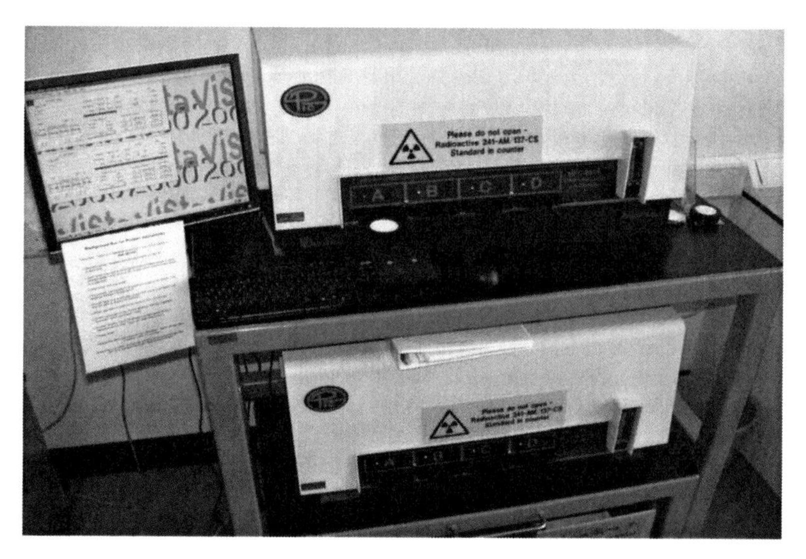

Figure 1 *Gas proportional counter*

The final procedure for the rapid gross alpha and beta method removes a whole step and can be summarised as follows:

1) A 100ml aliquot of acidified sample with bulking agent added is sulphated and evaporated.
2) The residue is then ignited at 350°C for 30 minutes.
3) The residue is then ground and planchetted.
4) The planchette is then counted on a gas proportional counter for a minimum for 30 minutes.

The resultant method achieved all the objectives set for the rapid radioactivity project. This method was then presented to the UK water laboratories mutual aid scheme (WLMAS) for use by the water industry in an emergency in 2003. Companies with radiochemistry capability validated the method using their own staff and instrumentation.

5 MUTUAL AID RADIOACTIVITY SUB-GROUP

The mutual aid radioactivity sub-group was established in 2006 and meets annually with participation from all UK water laboratories with radioactivity capabilities, LGC, Drinking Water Inspectorate (DWI), and other government agencies. The aims of the group are as follows:

- To facilitate the organisation of inter-comparison exercises
- Result sharing and interrogation.
- Promoting good practice
- Discussion of topics pertinent to the group including: training, proficiency schemes, conferences and seminars.
- Acting as a forum for the water industry to discuss issues and form opinions then speak with one voice.
- Lobby relevant bodies for assistance, clarification, modifications etc.

The establishment of the group has enhanced capability throughout the industry. Some collaborative exercise results are given in Tables 4 and 5. The results clearly indicate that the rapid method with a 30 min counting time is suitable for rapid screening to determine if the proposed limits 5 Bq/l for gross alpha and 30 Bq/l for gross beta have been exceeded. The results obtained by extending the counting time from 30 min to 16 hours (see Table 5) are not significantly different. A laboratory with a 10 channel counter and an adequate number of trained staff should be capable of analysing up to ~ 150 samples per day using the proposed rapid method. It is now proposed to include this rapid gross alpha/beta method in the next revision of the Standard Committee of Analysts (SCA) full scale method[10]

6 CONCLUSIONS

The development of the rapid alpha beta method has provided the UK water industry with a robust method for use in emergency situations. When used in such situations, if the results of analysis are below the Screening Levels, it can be presumed that they are below the Action Levels and no further isotopic analysis is required. The water can be used in the short term if necessary. If levels are above the Screening Levels further isotopic analysis is required to determine whether the Action Levels have been exceeded, and if this is the case contamination of the water must be reduced before it can be used for drinking.

A laboratory with a 10 channel counter and an adequate number of trained staff should be capable of analysing up to ~ 150 samples per day using the proposed rapid method with a 30 min count time. With the full scale method the maximum sample throughput is ~ 8 samples per day.

Table 4 *LEAP TEST 6 – Radioactivity Results 30 minutes (Spring 2010)*

Lab	Gross Alpha Bq/litre	Gross Beta Bq/litre	Sample Volume ml	Date & Time Analysis Started	Count Completed
001	2.49 2.08	10.34 10.57	91.1 88.3	18/05 – 10:10	18/05 – 23:00
002†	1.02 2.37	8.97 8.97	100	18/05 – 10:15	18/05 – 14:50
003	1.02 2.37	8.97 8.97	100	18/05 – 10:15	18/05 – 14:50
006	1.85 2.03	9.937 10.924	100	18/05 – 10:30	18/05 – 16:00
007†	1.02 2.37	8.97 8.97	100	18/05 – 10:15	18/05 – 14:50
009	2.067 2.871	9.396 8.822	100	18/05 – 10:15	18:05 – 16:19
010†	1.02 2.37	8.97 8.97	100	18/05 – 10:15	18/05 – 14:50
011	0.182* 0.373*	0.963 0.931	100	18/05 – 11:00	18/05 – 17:15
012	1.626 1.983	9.079 9.845	100	18/05 – 10:45	18/05 – 17:25
014	1.02 2.37	8.97 8.97	100	18/05 – 10:15	18/05 – 14:50
017	0.164 0.154	10.145 9.165	461.9	18/05 – 10:00	18/05 –10:30

† Indicates that laboratory sub-contracted its analysis.
* The laboratory analytical quality control action failed, the results would not normally be reported.

Table 5 *Mutual aid radioactivity sub-group Inter-comparison exercise (Sept 2009)*

Laboratory	Alpha rapid (30 mins) count time	Alpha full 16h Count time	Beta rapid (30 mins) count time	Beta full 16h Count time
1	3.79	3.24	4.1	3.76
	3.89	3.13	3.65	3.43
2	2.55	2.83	3.99	3.94
3	2.0511		2.7534	
4	3.193	3.041	3.788	3.934
	3.833	3.012	4.064	4.049
5	3.476	3.015	3.763	3.751
	3.375	3.326	3.926	3.593
6	3.757	3.186	4.326	4.311
7	2.97		3.76	
Mean	3.29		3.81	
S.D	0.61		0.42	
Spike Values	3.194 (Am241)	3.194 (Am241)	5.425 (Cs137)	5.425 (Cs137)

References

1. *COUNCIL DIRECTIVE 98/83/EC of 3 November 1998 on the quality of water intended for human consumption (OJ L 330, 5.12.1998, p. 32)*
2. *The Water Supply (Water Quality) Regulations* 2000 (No. 3184)
3. WHO (2004) *Guidelines for drinking-water quality, 3rd Ed, Vol 1. Recommendations*, Geneva World Health Organisation , ISBN 92 4 154638 7
4. NRPB (1994). *Guidance on restrictions on food and water following a radiological accident.* Doc. NRPB, 5, No.1. (London, HMSO).
5. CEC (1989) *Council Regulation (Euratom) No. 3954/87 laying down the maximum permitted levels of radioactive contamination of foodstuffs and feeding stuffs following a nuclear accident or any other case of radiological emergency.* Off. J. Eur. Commun., L371/11 (1987), amended by Council Regulation 2218/89. Off. J. Eur. Commun., L211/1.
6. Health Protection Agency, 2008. *Handbook for Assessing the Impact of a Radiological Incident on Levels of Radioactivity in Drinking Water and Risks to Operatives at Water Treatment Works.* [Online] (Updated 1 September 2008). Available at: http://www.hpa.org.uk/Publications/Radiation/HPARPDSeriesReports/HPARPD040/[Accessed 7th December 2010].
7. *Emergency Screening Levels for Gross Alpha and Beta Activities in Potable Water* NRPB-EA/3/2002
8. BS ISO 9696: 2007, *Measurement of gross alpha activity in non-saline water: Thick source method.*
   BS ISO 9697: 2008, *Measurement of gross beta activity in non-saline water: Thick source Method.*
9. National Compliance Assessment Service Technical report. *"Review of Alpha and Beta Blue Book Methods; Drinking Water Screening Levels.* E.A., NCAS/TR/2002/003 Feb.2002
10. *Measurement of Alpha and Beta Activity of Water and Sludge Samples. The Determination of Radon 222 and Radium 226. The Determination of Uranium (including General X ray Fluorescent Spectrometric Analysis)* Standing Committee of Analysts, 1985 ISBN 011751909X

A SCANDINAVIAN EMERGENCY FOR DRINKING WATER NETWORK CONTAMINATION: THE NOKIA CASE STUDY

I.T. Miettinen[1], O. Lepistö[2], T. Pitkänen[1], M. Kuusi[3], L. Maunula[4], J. Laine [3,5] and M-L. Hänninen [4]

[1]Water and Health Unit, National Institute for Health and Welfare, P.O. Box 95, FI-70101 Kuopio, Finland
[2] Department of Environmental Health Care, Pirkkala community, Suupantie 6, FI-33960, Pirkkala, Finland
[3]National Institute for Health and Welfare, Mannerheimintie 166, FI-00300, Helsinki, Finland
[4] Department of Food and Environmental Hygiene, Faculty of Veterinary Medicine, P. O. Box 66, FI-00014, University of Helsinki, Finland
[5]Tampere University Hospital, P.O.Box 2000, FI-33521 Tampere, Finland

1 INTRODUCTION

In 1997, a new notification system for waterborne outbreaks was launched in Finland. In this system, municipal health protection authorities have an obligation to notify national authorities of all suspected waterborne outbreaks. A notification of an outbreak has to be given as soon as possible after a suspicion of an outbreak linked to the quality of drinking water has came out i.e. before confirmative microbiological and chemical analyses of the quality of drinking water have been carried out. The notification report is actually an electronic questionnaire available on the Internet. The questionnaire has to be filled by municipal health protection authorities who in Finland are responsible for frequent surveillance of the quality of drinking water. The National Institute for Health and Welfare (THL) is involved with waterborne outbreaks by maintaining a national task group, which helps local authorities in technical, analytical and epidemiological problems associated with outbreaks. In 1998 – 2009, 67 outbreaks resulting in 27,200 illness cases have been registered (unpublished results). Outbreaks have typically been associated with the small ground water works serving less than 500 consumers. Noroviruses and *Campylobacter jejuni* have been the most common microbes causing the waterborne outbreaks in Finland.

2 THE OUTBREAK

The technical reason causing the massive contamination of drinking water distribution network at Nokia was the cross-connection between treated waste water and tap water pipelines at the waste water treatment plant in November-December 2007. The connection was open for two and half days resulting in intrusion of 450 m^3 of treated waste water into the drinking water distribution system. The treated waste water (waste water treatment included chemical precipitation and activated sludge treatment) has been used for at least two decades in the waste water plant for cleaning the surfaces and to dilute chemicals in the waste water treatment process. The waste water plant is located in the city centre of Nokia (30,000 inhabitants).

The first signals of the contamination were the customer complaints beginning on 28th November about abnormal drinking water quality. The complaints related to odd colour, taste, smell, and foam formation in the tap water. The first complaints were directed to the control room of the Nokia waterworks. They received several hundreds of complaints during the first few days of the contamination. No immediate actions were made because the personnel of the waterworks supposed that the reason for complaints about the water quality could be associated with loosened deposits. The deposits causing water quality issues were assumed to be derived from the changes in water flow direction in the distribution network due to an unusually intensive water intake.

The first cases of illness appeared on 30th November in the local health care centre. Symptoms included nausea, vomiting, diarrhoea and fever. Once information about water quality complaints, which seemed to be related to the gastroenteritis reached the local health inspector, the outbreak was finally recognized. [1] This finding finally initiated the mitigation actions to stop the outbreak. Very high counts of indicator bacteria, *E. coli* and intestinal enterococci, were identified from the first tap water samples taken from the network. Later, also *Campylobacter* sp., noroviruses, *Salmonella sp.*, and rotaviruses were detected both from tap water and human faecal samples. [2, 3] In addition, entero-, astro- , rota- and adenoviruses were also detected from water samples.[2] Later on also *Giardia* sp. was identified as a cause of human infections and cysts were also detected from one deposit sample taken from the drinking water distribution network in the beginning of December 2007 .[4]

3 MITIGATION ACTIONS

The personnel of the waterworks were able to locate the source of contamination within the same day as the contamination was confirmed. The connection between the wastewater line and tap water line was immediately cut and flushing of the network with clean water was started. Based on analytical measurements and hydraulic information, it was possible to define the contaminated area so that decontamination actions could be addressed on the contaminated area with 9,195 inhabitants. The distributed drinking water in Nokia city is normally disinfected with chlorine (approx. 0.3 mg/l). Among first mitigation actions chlorine concentration was first lifted up to 1.5 mg/l. Later it was raised up to 5 mg/l. Due to severe contamination of distribution network pipeline internal gauging (pigging) and air/water pulsed flushing were used to clean the main pipelines. Additionally, shock chlorination (10 mg/l, 24 h) was used to enhance cleaning of especially the connection lines and the indoor pipeline installations. Eighteen alternative drinking water service points (water tanks and distribution of bottled water) were also opened in the contaminated city area.

A boiling water advisory (5 min boiling time) of drinking water was ordered within the same day as the outbreak was finally recognized. Information on the contaminated water and boiling water advisory was spread using the local radio channels, Internet, TV and newspapers. Leaflets and loudspeaker cars were used to inform customers. Also, public centres like schools, food factories and hospitals were informed of the drinking water contamination to avoid use of contaminated (un-boiled) water as drinking water. Due to severe contamination of the network, boiling water advisory notices continued for almost three months. These were cancelled gradually in February 2008 after contaminated pipelines were cleaned and no pathogens, or indicator bacteria were found in the water samples taken after the cleaning procedures.

4 OUTCOMES OF THE OUTBREAK

A population survey was conducted in Nokia and the neighbouring town to find out the extent of the outbreak and to evaluate its consequences. The outcome of the study was that a total of 8,453 cases of gastroenteritis occurred during the outbreak, the excess number of illnesses being 6500. 1222 visits were made to the municipal health centre and nearly 200 consumers were treated in hospital.[1] Due to the contamination of the tap water, service points were organized to offer an alternative drinking water to the inhabitants of Nokia city. A total of 5,000,000 litres of tank water and 700,000 litres of bottled water were distributed for customers during the outbreak. Total costs of the outbreak including the mitigation actions, reimbursed hospital expenses, claims for damages, water analyses, and extra workload exceeded 4.6 million Euros.

5 LESSONS LEARNED

The largest waterborne outbreak in Finland for at least 50 years strongly affected the national water services. Unlike many other outbreaks which have occurred in Finland, several intestinal pathogens were isolated and identified as contaminating the water and causing the outbreak. This was also the first waterborne outbreak associated with drinking water contaminated by *Giardia* protozoa for over 15 years in Finland. [4] As in many (waterborne) outbreaks, the significance of the public communication was found to be extremely important. Earlier recognition and then rapidly informing the public of the outbreak could have limited substantially the extent of the outbreak. More transparent and efficient information, especially in the beginning of the outbreak, would have diminished the customers´ criticisms concerning the capability of authorities to manage the incident.

As a result of the outbreak, a national survey was launched by the Finnish Ministry of Social Affairs and Health to check and remove similar waste water and clean water pipeline connections, which allowed the Nokia outbreak to occur.

References

1. J. Laine, E. Huovinen, M.J. Virtanen, M. Snellman, J.Lumio, P. Ruutu, E. Kujansuu, R. Vuento, T. Pitkänen, I.T. Miettinen, J. Herrala, O. Lepistö, J.. Antonen, J.Helenius, M-L Hänninen, L. Maunula, J.Mustonen, M. Kuusi, M., and the Pirkanmaa Waterborne Outbreak Study Group. *Epidem. Infect.*, 2010, doi: 10.1017/S0950268810002141.
2. L. Maunula, P. Klemola, A. Kauppinen, K. Söderberg, T. Nguyen, T. Pitkänen, S. Kaijalainen, M. L. Simonen, I. T. Miettinen, M. Lappalainen, J. Laine, R. Vuento, M. Kuusi and M. Roivainen. *Food Environ. Virol.* 2009, **1** (1): 31-36.
3. T. Pitkänen, L. Maunula, M-L. Hänninen, M.-L., A. Siitonen, O. Lepistö, M. Mäkiranta, M. Kuusi, I.T. Miettinen. 201x, *In preparation.*
4. R. Rimhanen-Finne, M-L. Hänninen, R. Vuento, J. Laine, S.T. Jokiranta, M. Snellman, T. Pitkänen, I.T. Miettinen, and M. Kuusi. Scan. J. Infect. Dis., 2010, **42** (8), 613-619.

SENSORS AND WEBSERVICES FOR LAND & WATER MANAGEMENT

W. Boënne[1], X. Tang[1], N. Desmet[1], J. Schepens[1], P. Seuntjens[1,2]

[1] VITO, Environmental Modelling, Boeretang 200, 2400 Mol, Belgium
[2]UGent, Department of Soil Management, Coupure links 653, B-9000 Gent
Contact : wesley.boënne@vito.be

1 INTRODUCTION

Water resources are increasingly being put under pressure due to extensive growth of emerging economies, climate change and the increasing number and size of environmental incidents (e.g. oil spills, algal blooms). Decline of water quality is a growing concern with respect to public health and safety in many EU countries and developing economies throughout the world. This paper illustrates the application of in-situ (mobile) sensors for water quality (groundwater, surface water, waste water) and the integrated spatial data infrastructure as a management tool for protection of land and water resources.

Mobile sensors or sensor networks offer high temporal and spatial resolution monitoring data, using both direct and indirect measuring technology. Assimilation of relevant datasets supports science & policy makers understanding complex water systems. Data processing and modeling, using "smart software" allows the estimation of derived water quality parameters, data screening (avoid error data) and to set alarms (early warnings). The proposed system integrates (mobile) sensor platform services i.e. *in situ* measurement and visualization of water characteristics. Data are converted to useful information, usually visualized in a web client application.

In the described case study, a series of terrestrial and meteo sensors allows us to monitor subsurface flow and surface runoff in order to identify / quantify pathways of chemical transport from land to the river. These sensors combined with a meteorological station are connected to a fully automated solar-powered data logging system with wireless communication to a lab-based PC. Data are transferred to a database and subsequently information is visualised in a demo web application http://vastesensoren.rma.vito.be:8080.

2 APPROACH

This chapter describes the proposed integrated solution, which combines observation with relevant sensors for *in situ* measurement, spatial data infrastructure, modelling and (web based) visualisation of output towards end users.

The general architecture and the associated services are displayed in Figure 1 for an embedded sensor array, but these also apply to a mobile sensor platform or even an array of mobile platforms.

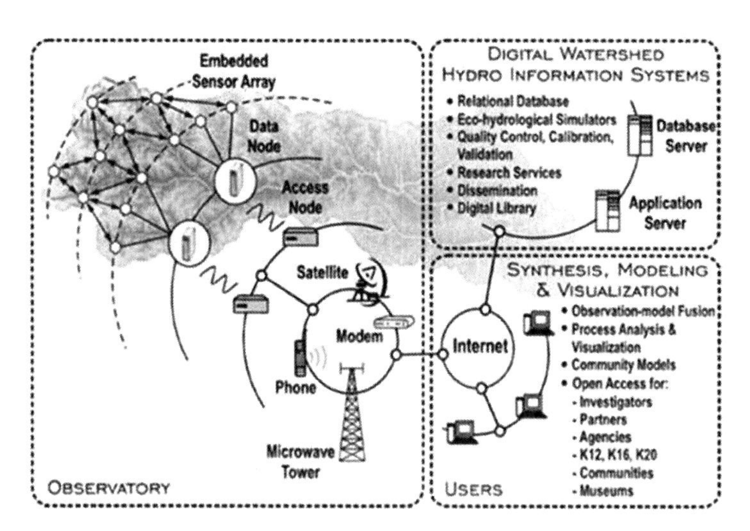

Figure 1 *The general architecture displays the observatory sensors on (mobile) platform(s), communication and spatial data infrastructure (e.g. a geo-server hosting geospatial databases) feeding models from which output will be forwarded visually to end users*

2.1 Observations

A series of terrestrial and meteo sensors (Figure 2) allows us to monitor subsurface flow and surface runoff in order to identify / quantify pathways of chemical transport from land to the river. These sensors combined with a meteorological station (Campbell Scientific) are connected to a fully automated solar-powered data logging system with wireless communication (GSM or GPRS) to a lab-based PC.

Sensors can serve water quality, - quantity, - current, bathymetrics and even navigation purposes on mobile sensor platforms (Figure 4). Prior to the purchase of the sensor equipment, an inventory of the available state of the art sensors was performed. Latest trends show a shift towards optical measurements.[1] The selected water quality sensors were validated comparing *in situ* measurements with lab analysis (VITO reference lab for Flanders for environmental analysis). These *in situ* sensors (Figure 3) included a multi parameter electrode (YSI 6600 V2), which covers a broad range of parameters and a spectral analyzer (S::CAN spectro::lyser), which proved indirect optical measurements can be a good proxy for the actual parameter.

Figure 2 *Terrestrial sensors (TDR, drop counting glass fibre wick, run off tipping buckets, tensiometers) and meteo station connected to a fully automated solar-powered data logging system with wireless communication to a lab-based PC*

Figure 3 *Multiparameter electrode (YSI 6600 V2) and S::CAN spectro::lyser*

The spectral sensor is easier to use and calibrate than *in situ* wet chemical analyzers and has excellent long-term stability. It can be used to measure different carbon structures, dissolved organic carbon and nitrates. Both sensors showed a linear response in the lab, within the measuring range specified by the supplier. However, for spectral field studies an additional "local" calibration, aside from the suppliers global calibration (river, waste water, drinking water, ...), is required in order to get a quantitative result, compensating for specific matrix effects. The ion selective electrodes on the multiparameter probe, used for estimating nutrients (nitrate, ammonium) and chlorides, need regular calibration as they are subjected to drift. Lab tests enabled us to compensate for matrix effects of chloride on nitrate measurements in brackish water.

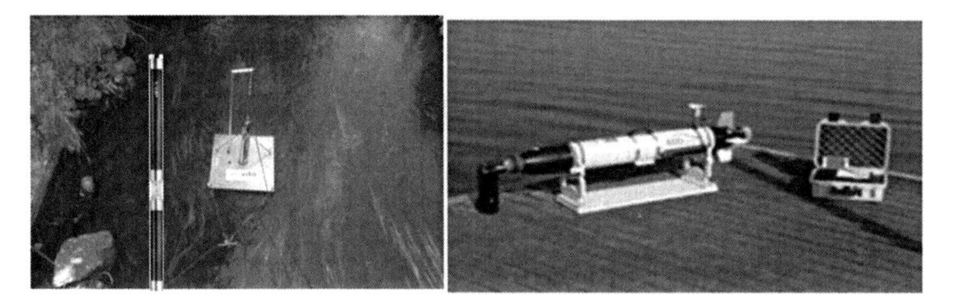

Figure 4 *Water sensor platforms: Float with multiparameterelectrode (YSI 6600 V2) and S::CAN spectro::lyser and an autonomous under water vehicle (Ecomapper, YSI) with water quality sensors in its nose*

These sensors can perform autonomous measurements at a high temporal resolution and report (near) real time. These sensors can map spatially as they will be deployed on an AUV (autonomous underwater vehicle) used for environmental mapping (e.g. buffer ponds), disaster management,... or other stakeholders needs.

2.2 Geospatial data infrastructure

What follows describes the technical outline of the integrated solution, the geospatial data infrastructure (GDI) and Sensor Web Enablement (SWE). Figure 5 illustrates the software architecture of the GDI.

2.2.1 Client application. The system will be accessible through a browser as a Rich Internet Application. It reflects the transition of specialized web applications from the simple thin-client model of a traditional web browser to a richer distributed-function model that behaves more like the desktop in a client/server model. Today, these richer user experiences are being implemented with technologies such as Java, AJAX and Flash, using standard internet and web protocols.

2.2.2 Service layer. The service layer defines the services that are available in the Rich Internet Application. Some part of the functionality of the services runs in the client application (browser). Most part of the functionality however will run at the backend in the server applications. A common service is the map view service, which allows for visualization of geo-referenced data.

2.2.3 Server Layer. The server layer is composed of 4 major components: web server, map publishing server and database server and catalog server.

The web server is Apache TomCat. It powers numerous large-scale, mission-critical web applications across a diverse range of industries and organizations. Basically, it's an implementation of the Java Servlet and JavaServer Pages technologies.

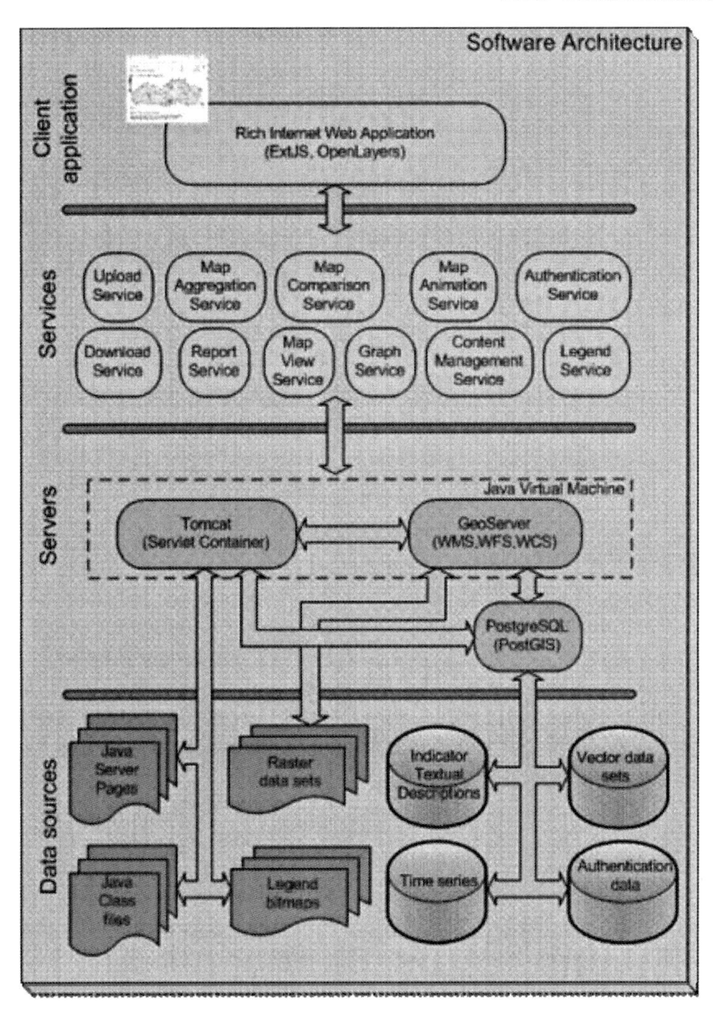

Figure 5 *Software architecture geospatial data infrastructure (GDI)*

As map publishing server, GeoServer will be used. GeoServer is an implementation of the OpenGIS Service Framework. This framework defines interfaces and protocols for collaborating geographical services facilitating the discovery, publishing and binding of geographical data. It is an open source software server written in Java that allows users to share and edit geospatial data. Designed for interoperability, it publishes data from any major spatial data source using open standards. GeoServer is the reference implementation of the Open Geospatial Consortium (OGC) Web Feature Service (WFS) and Web Coverage Service (WCS) standards, as well as a high performance certified compliant Web Map Service (WMS). The WebMap Service is the solution for the interoperable visualization of GIS-information over the Internet. The Web Feature Service aims at the distribution of geographic vector data over the Internet. It can also be used as the protocol for the manipulation of a geographical data set at a distance. The Web Coverage Service is the protocol for the open exchange of geographical raster data.

PostgreSQL will be used as database server (Figure 6). It is a powerful, open source object-relational database system. PostGIS adds support for geographic objects to the PostgreSQL object-relational database. In effect, PostGIS "spatially enables" the PostgreSQL server, allowing it to be used as a backend spatial database for geographic information systems. Currently PostGIS only supports vector data sets.

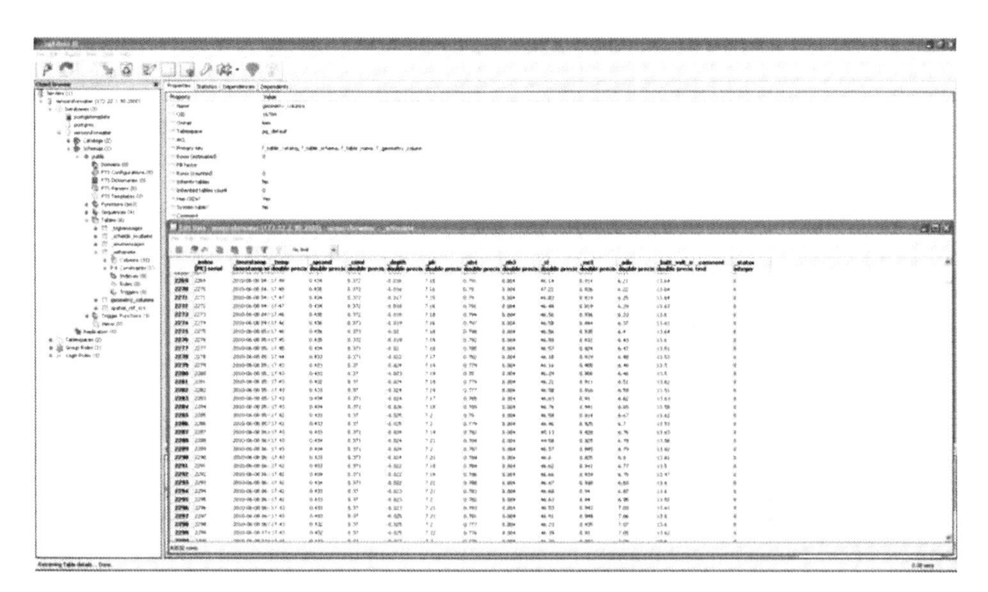

Figure 6 *PostgreSQL Tool*

As catalog server, GeoNetwork will be used. It is an open source web based geographic metadata catalog system supporting OGC Catalog Service for the Web (CSW 2.0). Catalog services support the ability to publish and search collections of descriptive information (metadata) for data and services. Metadata in catalogues represent resource characteristics that can be queried and presented for evaluation and further processing by both humans and software. Maps, including those derived from satellite imagery are effective communicational tools and play an important role in the work of decision makers.

2.2.4 Data source layer. The data source layer contains all the cartographic data (geodata) as well as time series data. Raster data will be stored in GeoTIFF format in the file system. Vector data will be stored in the database. In order to import this data in the database it should be delivered in ESRI Shapefile format . For map visualization, we also need to add styling information to the data store. This results in style language descriptor files (SLD's) and legend bitmaps.

2.2.5 Sensor Web Enablement. As spatial data infrastructures (SDI) are well established for building distributed applications in the geospatial domain, it is of special importance to link them with sensors and sensor data. Traditional Open Geospatial Consortium (OGC) services allow you to request sensor data, but only in a limited manner. But a generic framework for sensor data integration into SDIs was missing. Thus it was obvious to extend the SDI specifications by a framework for integrating sensors into SDIs. Therefore the OGC founded the Sensor Web Enablement (SWE) initiative which is developing standards for access to and control of sensors and sensor networks via the Internet. The goal of SWE is to enable all types of internet-accessible sensors to be accessible and, where applicable, controllable via the world wide web.

2.3 Processing, modelling and webservices

As explained above, once deployed *in situ* data will be communicated to a ground station and stored on a geoserver hosting a geospatial database. Using "smart software", data will be screened in order to exclude non sense data and appropriate e-mail or SMS alerts are diverted to relevant persons involved (policy - decision makers, water managers, ...). Models will process and convert complex data into relevant information to the end user.

There are a number of approaches, particularly data assimilation techniques that can foster improved understanding and management by integrating atmospheric, terrestrial, and oceanic/aquatic data acquired from *in situ* and remote assets into a common analysis framework. Through explicit data assimilation methods, models can integrate a wide range of routine observations from *in situ* or remote sensing platforms, allow the use of imperfect or proxy measurements, help fill gaps in time or incomplete coverage, and provide a quantification of errors. Through inverse models, observations can be converted into the desired quantities, such as sources and sinks or key driving parameters that may not be directly measurable or are unobservable at a particular scale. In addition, through coupling and nesting, models can encompass the necessarily wide range of spatial and temporal scales to quantify sources, sinks, and fluxes.

The prototype integration software architecture for different sensors, sensor networks or mobile platforms, such as AUVs, UAVs (Unmanned Aerial Vehicles), cars, bikes, people etc... has already been established and is displayed in Figure 4.

As the information will be visualized on a web client application (Figure 7), end users do have global access to (near) real time data finally resulting in economical and/or socio-ecological benefit.

3 CASE STUDY: TERRESTRIAL SENSORS

The following terrestrial sensor setup assists multiple end users of water: the agricultural sector, water managers and drinking water companies. Irrigation management and information on the nutritional status are relevant to farmers. Water managers gain insight in water levels and water quality status. Drinking water companies track diffuse micropollutants (e.g. plant protection products) and nutrients in soil, groundwater and surface water.[2]

Pollutants can be transported to groundwater by leaching and to river systems by macropore flow in tile-drained fields and surface runoff in sloped areas, but also due to point sources such as urban drainage, effluents of waste water treatment plants, and inappropriate agricultural practices. Our study focuses on a fully automated field setup for quantitatively identifying fast flow processes through different preferential pathways in an agricultural soil towards a neighbouring river.[3]

Field plots at three locations along a representative hillslope of an agriculture-dominated remote watershed at Nil-Saint-Martin in Belgium have been setup to continuously monitor surface runoff, soil moisture and water potential, and to collect subsurface flow water (Figure 8). Runoff tipping bucket equipment, TDR probes, tensiometers, drop counting glassfibre wick samplers and a meteorological station were connected to a fully automated solar-powered datalogging system with wireless communication to a lab-based PC. Data are transferred to a database and subsequently visualised in a demo web application (Figure 9) at http://vastesensoren.rma.vito.be:8080/ .

Figure 7 *End user web client application for bathymetrics*

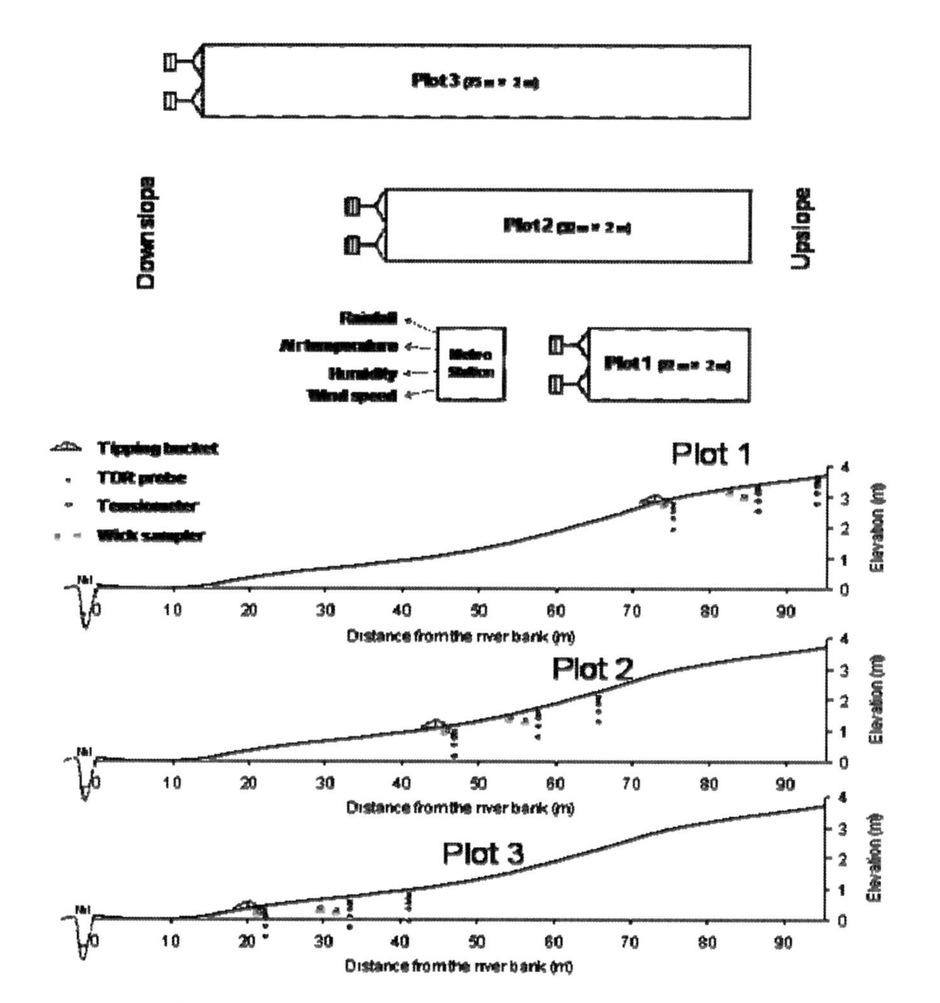

Figure 8 *Layout of field plots and experimental sensor setup*

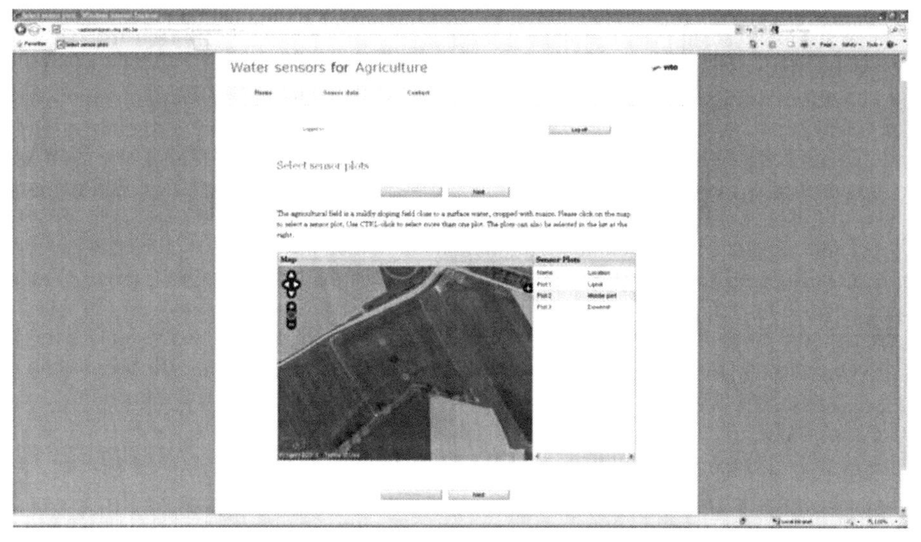

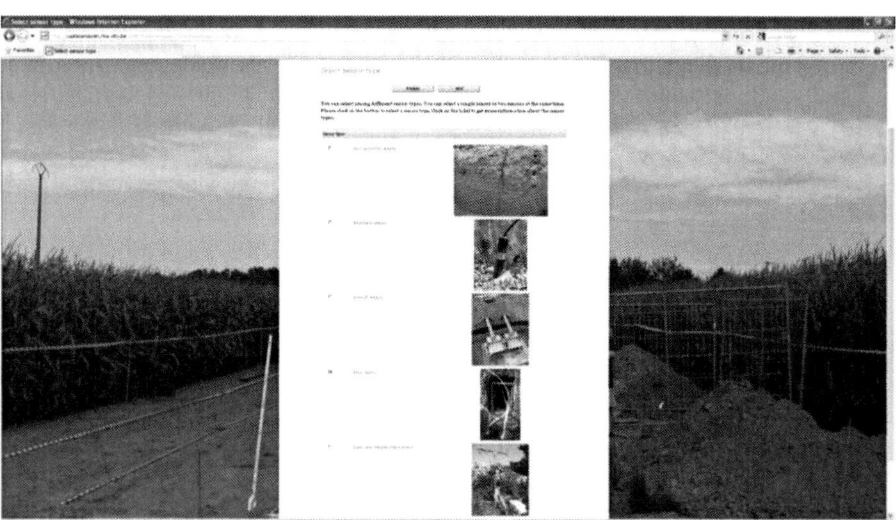

Figure 9 *Selection of field plots, sensor(s), timeseries and visualization of graphs with an option to filter or average data*

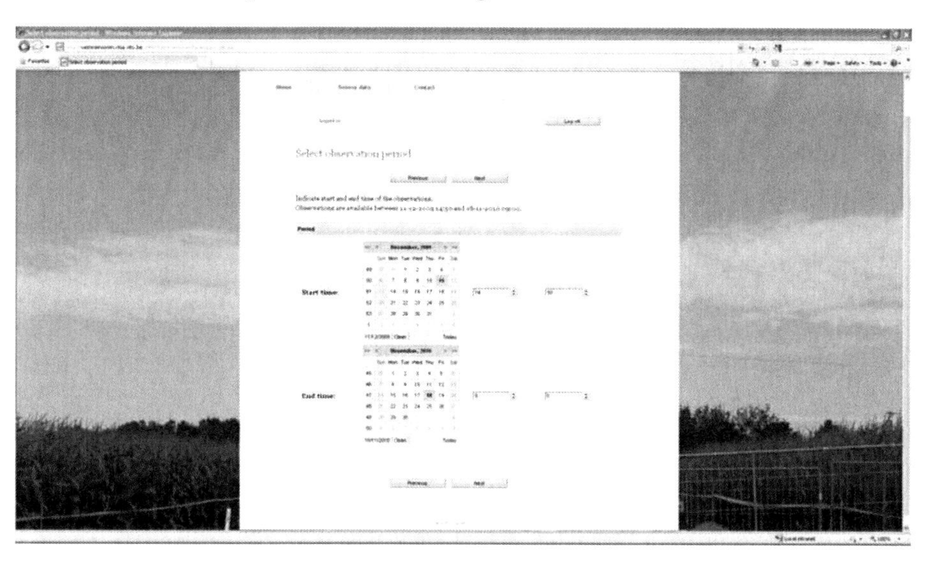

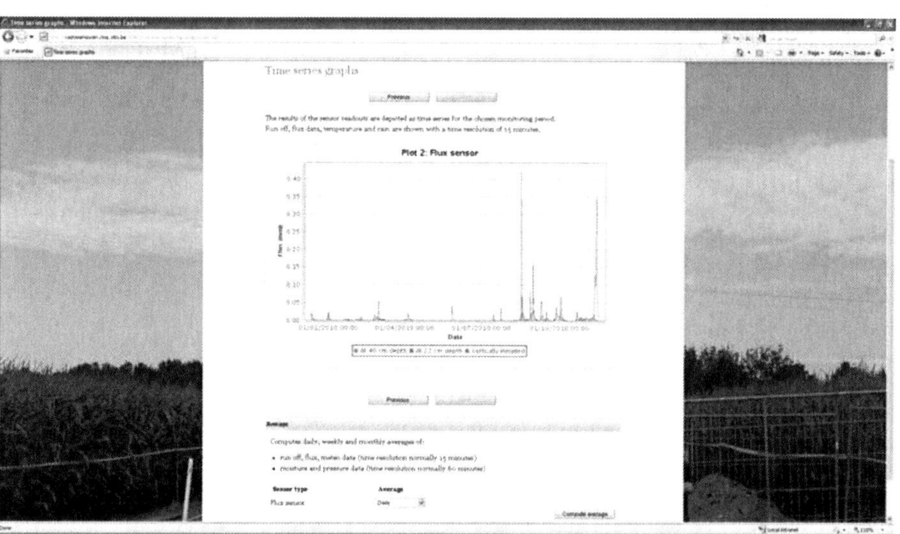

Figure 9 *continued*

An automated field monitoring system provides continuous observations, which are crucial for a better understanding of tempo-spatial characteristics of various flow and pollutant transport processes on (sloping) farmland toward water bodies. It provides also more reliable field data for the calibration of soil hydraulic parameters. Lateral flow can occur at the depth of 15-25 cm when the bulk soil moisture is relatively high and the relatively compacted layer at 29-33 cm is nearly saturated, typically in the late winter monitoring period with low temperature and snowfall events. Apparently, the vertical infiltration accounted for the major part of the rainwater for all the three monitoring periods, indicating that soluble, persistent chemical compounds may migrate to the groundwater via leaching.

Long term monitoring for multiple years, especially event-based monitoring during stormflow events, and modelling effort involving a better strategy for parameter optimization and model calibration using field data, can provide insight into the effects of climate change on hydrologic processes and the transport and fate of pollutants.

4 CONCLUSIONS

The VITO project illustrates the application of *in-situ* sensors for water quantity (rainfall, run off, recharge, ...) and quality (groundwater, surface water, waste water) as a tool for protection of land and water resources.

Mobile sensors or sensor networks offer high temporal and spatial resolution monitoring data, using both direct and indirect measuring technology.

Assimilation of relevant datasets supports science & policy makers understanding complex water systems. Data processing and modeling, using "smart software" allows us the estimation of water quality parameters, data screening (avoid error data) and setting alarms (early warnings). Data are converted to useful information, which prevents ever growing and unstructured data cemeteries.

Wireless communication ensures sensor data (and metadata) collection in the VITO data centre. Subsequent data processing steps can be configured by a battery of relevant models. In this respect, different areas are to be covered: environmental monitoring, hydrographic survey, freshwater mapping, rapid environmental assessments, sensor developments, early warning systems, disaster management, extreme events – climate change.

A generic web based approach, based on Open Geospatial Consortium (OGC) standards and the INSPIRE guidelines, ensures the interface between sensor, geospatial database and the end user.

References

1 Zielinski, O., Busch, J. A., Cembella, A. D., Daly, K. L., Engelbrektsson, J., Hannides, A. K., and Schmidt, H.: Detecting marine hazardous substances and organisms: sensors for pollutants, toxins, and pathogens, *Ocean Sci. Discuss.*, 6, 953-1005, doi:10.5194/osd-6-953-2009, 2009.

2 Holvoet, KMA; Seuntjens, P; Vanrolleghem, PA. 2007. Monitoring and modelling pesticide fate in surface waters at the catchment scale, *Ecological Modelling* 209 (1): 53-64.

3 Tang X.Y., Seuntjens P., Cornelis W. M., Boënne W, Verbist K., and Van Hoey S: Preferential pathways of pollutants via overland and subsurface flow in an agriculture-dominated watershed: field monitoring and evaluation, *Journal of Environmental Monitoring*, submitted for publication.

GRAPH DECOMPOSITION AS OPERATIVE TOOL IN HYDRAULIC SYSTEM ANALYSIS – SECURITY ASPECTS

J. W. Deuerlein[1] and A. Wolters[1]

[1]3S Consult GmbH, Osteriede 8 – 10, 30827 Garbsen, Germany

1 INTRODUCTION

Since the attack against the World Trade Centre in 9/11 the security of drinking water supply systems against deliberate contamination with chemical, microbiological, radiological and nuclear substances (CBRN) has been studied intensively by researchers all over the world. Water distribution systems of cities in industrialized countries are grown over decades and consist of different assets like water treatment plants, storage tanks, pumps, valves and the distribution pipes. The pipe networks extend over considerable distances, often more than 1000 km. Due to its distributed nature the pipe system is especially vulnerable against intentional contamination.

Although there exist a number of highly developed hydraulic simulation engines that enable the calculation of flows, pressures and mass transport through the system there are many open questions as to prevention of attacks, detection of contaminants and development of emergency plans. At the center of research activities in the field of hydraulic network modeling has been the consideration of the development of sensor networks as early warning systems for contamination events. A large number of researchers from all over the world have tackled this complex issue mainly by formulating and solving a mathematical optimization problem. In 2006, the Battle of the Water Sensor Networks (BWSN) was undertaken as part of the 8th WDSA conference in Cincinnati[1].

Despite the existence of highly developed optimization tools for sensor placement like TEVA-SPOT[2], which that has been developed and published by the US-EPA, there are still open questions and unsolved problems. Even well-planned contaminant warning systems with a large number of well-positioned sensors cannot avoid that a certain time span exists between intrusion and detection or in the worst case even non-detections. Another important issue is that the algorithms for sensor placement are based on assumptions (e.g. input mass, duration). Variations of these assumptions can lead to different results.

As a consequence it is very important for both the planning engineer and the operating engineer to be aware of the limitations of the algorithms and to have a good understanding of the distribution system. Very detailed all pipe simulation models have reached a remarkable level of detail and support this task. However, the increasing level of detail of the hydraulic simulation models also makes it more and more difficult for the engineer to keep track of the actual state of the system, the connectivity components and their interaction. For example, in the case of an emergency, it is not straightforward identifying the correct valve closures for isolating the contaminant. In addition to existing hydraulic simulation packages, tools are required that allow for both simplifying the network as much as possible and going into detail where necessary.

In the first part of this paper, a decomposition concept of a general supply network graph is presented having different applications in hydraulic network modelling. Firstly, it improves the modellers' knowledge of the network. Secondly, it supports all kind of network calculations by simplifying the underlying graph structure. Often parts of the network graph can be identified that are not affected by parameter changes in other parts. Recalculations can be tailored to the minimal necessary subgraphs.

In the second part, risk analysis of a water supply network against intentional contamination with CBRN- substances (CBRN; Chemical, Biological, Radio Nuclear) is presented, which is based on the decomposition of the network graph. The outcome of the risk analysis will be used at the end of the paper for the development of prevention measures. Structural measures for reducing network vulnerability are distinguished from measures for improved network control. The latter to the main part concerns the use of on-line-hydraulic simulation models that are driven by real-time-data from distributed measurements in the system.

2 DECOMPOSITION CONCEPT

The decomposition of a water supply network graph into its connectivity components is based on linear graph theory[3]. It can be used for getting a better understanding of the network topology as well as operative decision support system. The decomposition procedure can be summarized as follows[4]:

In the first step all maximal connected components of the network graph are identified by using a breath first search (BFS). The definition of component includes that a path exists from each junction of the component to all other junctions. An important property of such a maximal connected component is that the possible spread of contamination is restricted to the component of its intrusion.

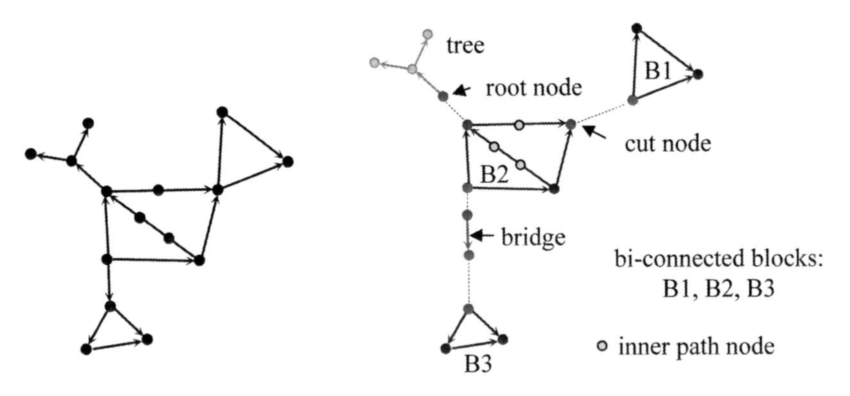

Figure 1 *Original network graph* **Figure 2** *Network graph after decomposition*

The components are further subdivided into looped and branched subgraphs. The looped blocks are interconnected by so called bridge components. The outer branched subgraph is called forest. The forest consists of trees that are connected to bridges or looped blocks at their root node. In contrast to the looped blocks with the possibility of changing flow directions the pipe flows of the branched subgraph are unidirectional. This is an important observation since it can be used for the development of prevention measures like installation of back flow prevention valves that avoid that contaminated water is pumped by a terrorist against the normal direction of flow.

Figure 1 shows an artificial network graph that is decomposed into its subgraphs in Figure 2. The network graph is connected; hence it consists of one maximal connected component only. The looped subgraph is composed of three blocks B1, B2, B3 that are connected by a bridge link (B2 and B3) and a cut node (B1 and B2), respectively. At block B2 one tree is connected being the only tree of the forest of the network graph.

As an example for the application of graph decomposition in risk analysis assume that the supply of the total system is in Block B1. Then, the flow in the bridge is always from B2 to B3. If water was pumped into the system in block B3 the contamination would be either restricted to this block or the flow direction of the bridge pipe had to be reversed, which can be avoided by installing non-return valves. By the same measures the tree could by unidirectionally separated from the rest of the system. As result the network is subdivided into smaller portions. In case of an intrusion accident the contaminant can be prevented from spreading over larger parts of the system.

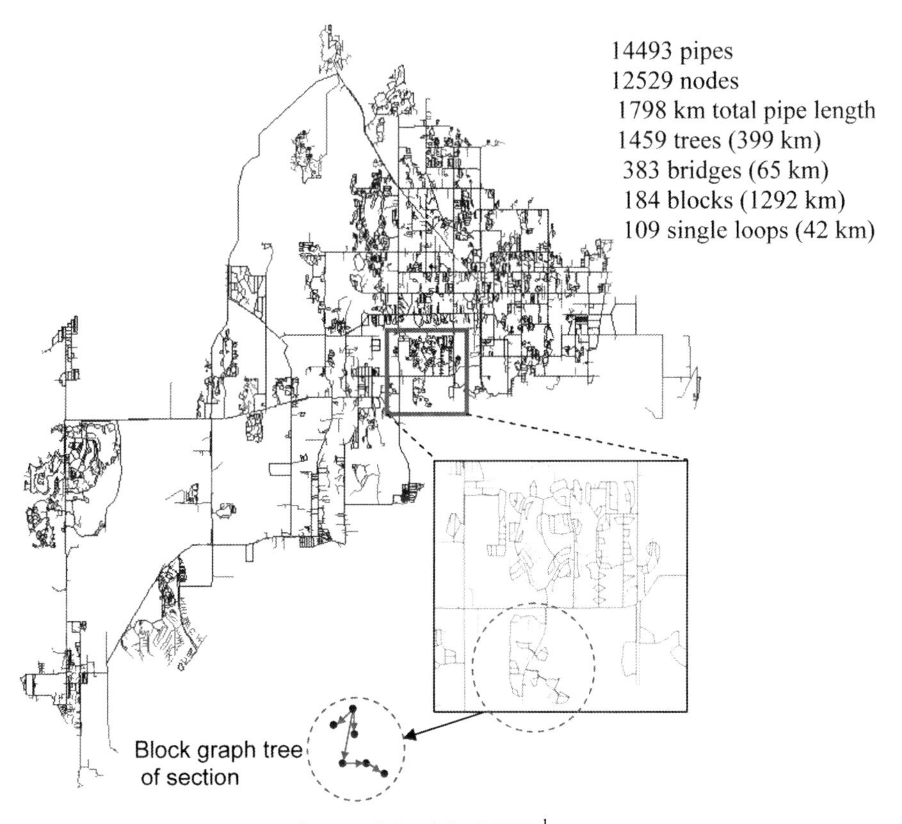

Figure 3 *Decomposition of network 2 of the BWSN*[1]

The knowledge of the graph components that result from the graph decomposition model has many applications in water network security: getting a very simplified view on the system by considering the block graph tree, distinguishing of highly endangered network parts (looped blocks) from less critical (tree structure), real time valve control for isolation of contaminants by network separation, identification of hydrants for flushing, optimal sensor allocation and last but not least prevention by developing less critical network structures.

Used as operative tool the decomposition algorithm enables the modeller to check the effect of decisions, like valve closures, on the network topology immediately without the need for time consuming hydraulic simulation runs. In addition only that network component must be considered where the attack has been detected. That simplifies the calculations and increases clarity.

Figure 3 shows the result of graph decomposition of a larger network: the example network 2 from the Battle of the Water Sensor Networks[2].

3 RISK ANALYSIS OF WATER SUPPLY NETWORKS

3.1 Definition of risk

In this section the risk of different locations of a water distribution system as to intentional contamination with CBRN- substances is studied. Possible intrusion scenarios are pumping of contaminated water into the system via hydrants or house connections and the contamination of a storage tank. In general, the risk of an event is defined as the product of the probability of occurrence of the event and its impact[5]:

$$R_i = P(E_i)\cdot \int_{t_0}^{t_1}\sum_{j=1}^{n_S} Q_{i,j}(t)dt$$

In our case the impact is estimated by the volume of contaminated water. R_i denotes the risk of intrusion at node i, $P(E_i)$ is the probability of occurrence, $Q_{i,j}(t)$ denotes the flow through pipe j leaving node i, t_0 and t_1 are the start time and the end time of intrusion and finally n_S is the number of pipe flows leaving node i.

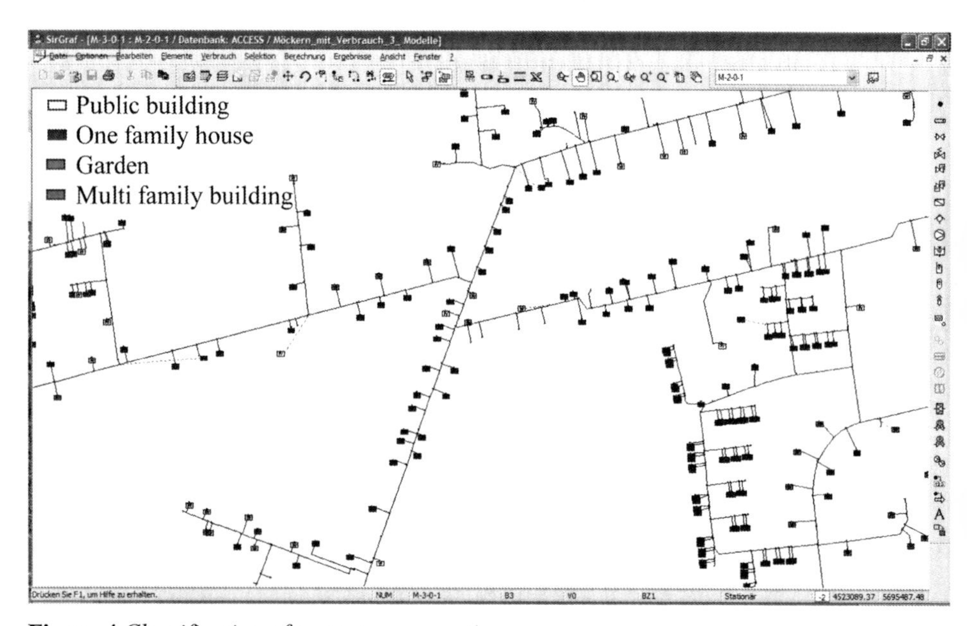

Figure 4 *Classification of customer connections*

Whereas the impact can be calculated by use of an hydraulic simulation model the estimation of the probability of occurrence ($P(E_i)$) of an intrusion at location i is not straightforward. Here, an all pipe hydraulic simulation model (Figure 4) has been used that includes information about different customer types. The relative probability $P(E_i)$ of an event i is calculated based on the following assumptions.

3.2 Assumptions

For simplification it is assumed that $P(E)$ solely depends on the following three probabilities:

- Detection likelihood of the threat (P_D)
- Accessibility of intrusion point (P_{Acc})
- Attractiveness of target (P_{Att})

The probability that an intrusion in a public building or in a multi family house is detected is much bigger than it is for an intrusion from a one family house or a garden. The accessibility reflects how difficult it is for the terrorist to bring the substance into the pipe system. The reason for considering the attractiveness of the target is that it is assumed that the terrorist might try attacking a special building like a barracks, a hospital or the like from a location in the neighbourhood of the target. This selection of probabilities that determine together the probability of occurrence does not claim completeness. It only serves as example for demonstrating the approach.

3.3 Example

Table 1 *Probabilities of different intrusion locations*

Customer type	P_D	P_{Acc}	P_{Att}	$P(E)$
Public building	0,8	0,5	0,8	0,080
One family house	0,1	0,7	0,8	0,504
Garden	0,15	0,9	0,8	0,612
Multi family house	0,8	0,4	0,9	0,072

3.4. Risk map

For every possible intrusion scenario the risk is calculated and summarized in a risk map (Figure 5). Please note that the computational effort highly depends on the assumption about the input flow of dissolved contaminant. Let us first assume that the input flow volume is very small in comparison to the water consumed in the network. Then, it is obvious that the dosage of contaminant does not change the flow conditions within the network. Therefore only one hydraulic simulation run has to be done in advance. The impact of any location can be calculated based on the results of the single simulation run. In contrast high input flows of dissolved contaminant can affect the flow distribution of the system heavily. Therefore it is necessary that for those intrusion scenarios the network hydraulic is recalculated resulting in much longer calculation time for the risk analysis.

3.5. Observations

High risk locations are one family house connections, with their comparably high relative probability of occurrence, that are directly connected to a main distribution line (high impact). Figure 5 shows the results of the risk analysis of an existing water supply network. The location with the highest risk is the node, where a number of houses are directly connected to a main distribution pipe.

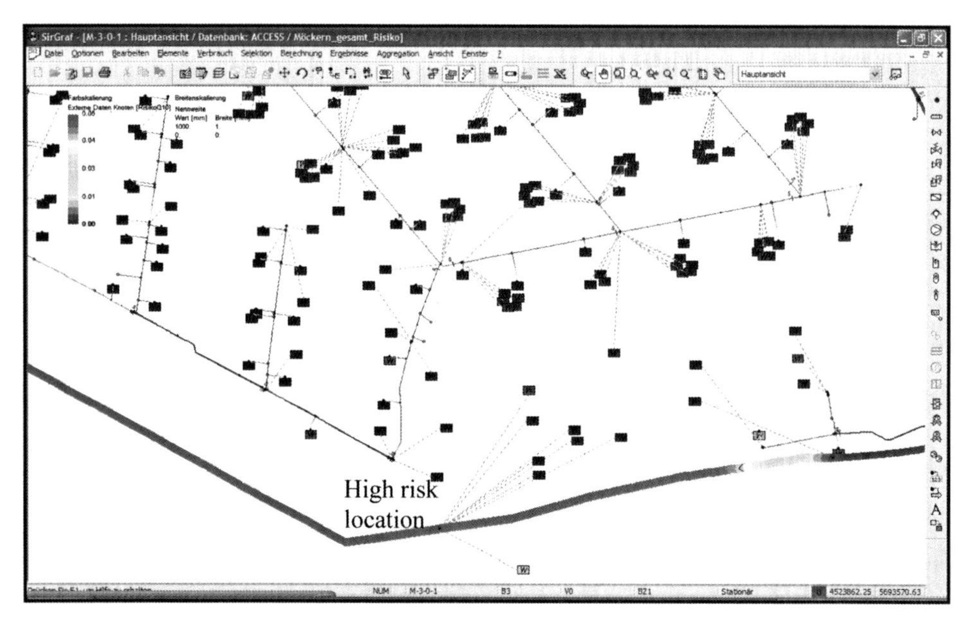

Figure 5 *Risk map*

The impact of an intrusion within the tree sub-graph is highly dependent on the input flow rate of contaminated water, which can be explained as follows.

In a tree subgraph under normal conditions the flow direction is unidirectional from the root to the leaves. If in case of an attack enough water is pumped into the system the flow direction may be changed. In this case it is possible that not only the customers downstream of the input location are affected but also a much bigger part of the network. One threshold value is the pipe volume between the input location and the root of the tree subgraph. If the terrorist could arrange replacing that water volume by contaminated water he could also reach a much more dangerous part of the network: the root node that connects the tree-subgraph to the superior supply network. As a consequence the contamination is no more local but can be spread over much larger network parts depending on the flow at the root node. As a result of the strong dilution at the root node mixing the much smaller reversed flow from the tree subnetwork with the water in the transport pipeline the initial concentration and toxicity of the substance plays also an important role. From this example it is clear that the estimation of the impact of an intrusion scenario is not straight forward and highly dependent on assumptions like mass input, toxicity etc. Therefore monitoring of the network should, in addition to quality sensors, also include the current hydraulic state of the network. This can be arranged by real-time-measurements in combination with hydraulic network simulation (on-line-simulation).

Another example that highlights the importance of considering real time data is the estimation of the spread of contamination within the looped part which depends on the current flows that in turn are a function of actual demands. Different demand distributions result in different flow patterns, and often flow directions change within the looped part. In case of an emergency, measures that are based on off-line hydraulic simulation models can result in dangerous misinterpretations.

4 PREVENTION MEASURES AND MONITORING

4.1. Prevention by unidirectional separation

The high risk resulting from the large number of house connections can be diminished by replacement or substitution of back flow preventers within a house connection by non-return valves that are located at public ground and cannot easily be accessed by an actor (see Figures 7 and 8).

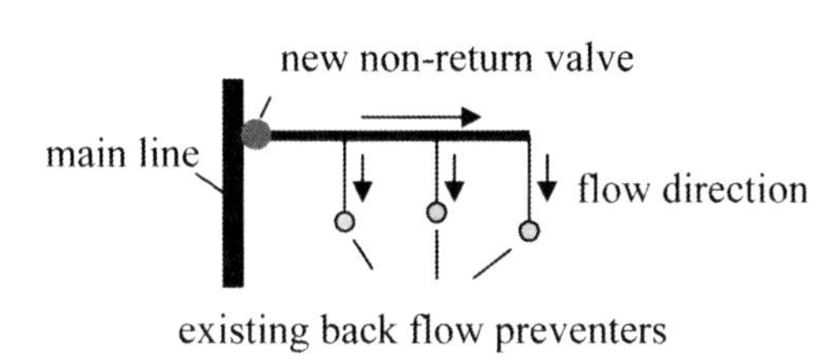

Figure 7 *Prevention by unidirectional separation*

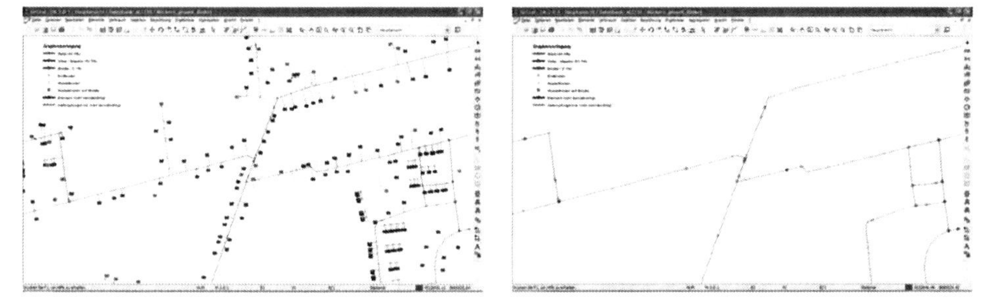

Figure 8 *Effect of unidirectional separation*

Shortcomings are that this is only a medium to long-term scale measure and the additional maintenance of the new fittings causing additional cost for the supply utility. As a consequence there is need for intelligent sensor networks in order to reduce network vulnerability. The intelligence can be gained from combination with on-line-hydraulic modeling.

4.2 On-line hydraulic simulation

Nowadays, Geo Information Systems (GIS) include detailed information about the pipe network. In addition, SCADA systems (SCADA: Supervisory Control And Data Acquisition) provide actual data of network hydraulics and water quality. Information about customer data and demands is available in billing systems and databases (Figure 9).

Often these systems are used and maintained separately causing redundancies and sources of error. Additional benefit could be gained if all the information was lumped. Particularly for increasing the security against contamination of the drinking water the use of all data available is essential. The billing system delivers, in addition to the yearly water consumption that is used for the calculation of nodal demands, also the kind of customer connection that, as explained above, can be used for the estimation of risk.

Figure 9 *Components of IT-Infrastructure at water supply utilities*

SCADA provides real-time data of currently measured values of pressure, flow and water quality at selected locations of the real physical system. However, real-time data is limited to only few measurement locations and mostly not linked to the network data. The shortcomings of the existing practice of running three separate software systems can be avoided by adding a fourth component: on-line hydraulic simulation.

Simulation puts all input data in a consistent model context (Figure 10). GIS and billing system are used for the implementation of the hydraulic model. In real-time mode the hydraulic on-line simulation model is driven by actual measurement data of the SCADA system. Some measurements serve as boundary conditions for the hydraulic solver (e.g. actual demands, pump and valve operations, tank water levels). Other data are used for the comparison of measured and calculated values (e.g. nodal heads, flows, water quality). If the difference between expected (calculated) and measured values is above a given threshold an alarm is generated automatically. One of the most important benefits is that, in comparison to conventional network control solely by SCADA systems, the entire network is considered including the interaction of its components. Whereas SCADA systems provide only local control mechanisms with on-line hydraulic simulation the network observation is globalized.

Figure 10 *Union of Billing-, GIS- and SCADA-data within on-line hydraulic simulation*

In case of an intrusion of contaminant the detailed knowledge of the current operational state of the entire system is required for the development of emergency plans and counter-measures. Only on-line hydraulic simulation can fill this gap:

- Without Simulation neither „Smart Grids" nor „Intelligent Networks"
- Monitoring, Understanding, Acting - a common buzz phrase sequence
- Understanding: Not possible without Simulation Models
- Monitoring + Understanding = Simulation + Simulation Result Analysis
- Acting: Look Ahead and What If Simulations secure optimized operating
- Dual benefit under "normal" conditions

4.3 Sensor placement

On-line hydraulic simulation requires sensors that are spread over the distribution system. Finding the most efficient sensor design by use of mathematical optimization belongs to the class of NP-hard problems. As a consequence, methods for reducing the size of the problem without loss of accuracy are needed. Graph decomposition provides those methods that can be combined with existing software tools for sensor allocation like TEVA-SPOT[2]:

- Only relevant intrusion scenarios and locations for sensors can be identified in a preliminary analysis.
- In all problems where a huge number of different calculations with small changes of boundary conditions is required graph decomposition and topological considerations can drastically reduce calculation time by selection of affected subgraphs.

5 CONCLUSIONS

Detailed hydraulic simulation models are required for a reliable estimation of risk. The more the number of features increases the more important are efficient tools for network simplification that allow different views on the network. The graph decomposition has been presented as example for such an approach.

Simulation of different intrusion scenarios for a real water supply network has shown that the input flow rate of contaminant intrusion is an important parameter for the estimation of impact, especially if the intrusion is in the branched network part. The risk abruptly increases when the input flow rate is big enough to reverse the flow in the branched subgraph and to reach the superior supply system. As a counter measure the reduction of network vulnerability as to intrusion of contaminant is possible by unidirectional separation of network parts (addition of non-return valves). Appropriate locations for the installation of those valves can be determined by graph decomposition. In general, separation of transmission and distribution is essential for reducing network vulnerability.

In our opinion one important shortcoming of existing methods for contaminant detection is that the mathematical models are all based on off-line simulations. However, the applicability and reliability of contaminant warning systems combined with hydraulic simulation models for early warning systems is crucial to real-time modeling. The real impact of a contaminant injection including the geographical coverage and the number of exposed people are all affected by the current water demand load and operation of the system. Off-line hydraulic solvers are not applicable as operational tools because results generated by hydraulic calculations may strongly distinguish from the actual state of the physical system. On-line-simulation tools are required that are capable of capturing the actual state of the system in seconds. Source identification and quick decisions on emergency measures like isolation of contamination, notification of population, flushing of contaminated pipes can only be well-directed if this information is provided by a process accompanying hydraulic solutions that always has the ability to reflect the current hydraulic state of the system with all its changing boundaries.

Consideration of risk in combination with the decomposition of the network graph can highly reduce problem size in sensor placement. The decomposition of the network graph also explains non-detections in existing sensor placement algorithms like TEVA-SPOT and can be used for the reduction of the number of intrusion points and the simplification of the problem by aggregation. The detection of all possible incidents in the branched forest subgraph of a real network is impossible in practice since a huge number of sensors would be needed. For the mathematical model the forest should be excluded. A further simplification can be reached by considering looped blocks separately and reducing the number of possible sensor locations to the path element nodes (nodes that connect three or more pipes) of the graph. If required a local analysis of a path element (sequence of pipes between two path element nodes) can be added.

The application of on-line simulation as operative tool can highly increase network security. However, that imperatively requires that the processor time needed for the calculations and preparation of results is far less than the time interval between two measurements (e.g. three minutes). In addition to process accompanying calculations "What-If"-"Look-Ahead"-scenarios play an important role. As a consequence, future research includes the development of faster hydraulic solvers, possibly by use of graph decomposition, and the efficient processing, interchange and presentation of the huge amount of data stemming from different data sources.

References

1 A. Ostfeld, J. G. Uber, E. Salomons, J. W. Berry, W. E. Hart, C. A. Phillips, J.-P. Watson, G. Dorini, P. Jonkergouw, Z. Kapelan, F. di Pierro, S.-T. Khu, D. Savic, D. Eliades, M. Polycarpou, S. R. Ghimire, B. D. Barkdoll, R. Gueli, J. J. Huang, E. A. McBean, W. James, A. Krause, J. Leskovec, S. Isovitsch, J. Xu, C. Guestrin, J. VanBriesen, M. Small, P. Fischbeck, A. Preis, M. Propato, O. Piller, G. B. Trachtman, Z. Y. Wu and T. Walski: The Battle of the Water Sensor Networks (BWSN): A Design Challenge for Engineers and Algorithms; *Journal of Water Resources Planning and Management*, 2008, 134 (6), 556 – 568.

2 J. Berry, E. Boman, L. A. Riesen, W. E. Hart, C. A. Phillips, J.-P. Watson and R. Murray. *User's Manual TEVA-SPOT Toolkit Version 2.3.* Report EPA 600/R – 08/ 041A National Homeland Security Research Center, Office of Research and Development, 2009. Download, April 13th 2010 at: www.epa.gov/NHSRC/pubs/600r08041a.pdf.

3 R. Diestel. *Graph Theory.* Springer Verlag Heidelberg, New York, 3rd Edition, 2005.

4 J. W. Deuerlein. Decomposition model of a general water supply network graph. *ASCE Journal of Hydraulic Engineering,* 2008, 134(6), 822 - 832.

5 B. Jiang. Risk Management and the Office of Homeland Security's Antiterrorism Task. *The Online Journal of Peace and Conflict Resolution* 4.2: 30 – 36, 2002.

Acknowledgment

This research is part of the multi-partner research project "STATuS" that is funded by the German Federal Ministry for Education and Research.

EFFICACY OF FREE CHLORINE AGAINST WATER BIOFILMS AND SPORES OF *PENICILLIUM BREVICOMPACTUM*

V.M. de Siqueira and N. Lima

IBB - Institute for Biotechnology and Bioengineering, Centre for Biological Engineering, University of Minho, Campus Gualtar, 4710-057 Braga, Portugal

1 INTRODUCTION

Biofilm is a complex community of microbes (bacteria, protozoa, filamentous fungi, yeasts and other microorganisms), organic and inorganic material accumulated amidst a microbial produced extracellular polymeric substances (EPS) attached to a surface.[1]

The structural and phenotypic changes associated with the development of a mature biofilm aid microbial cells in numerous aspects of their life cycles. One frequently measurable change in the phenotype of cells in a biofilm, when compared to their planktonic counterparts, is the significant increase tolerance to chemical, biological or physical stresses.[2,3] Other benefits may include tenacious attachment to surfaces, colonization of host tissues, expression or enhancement of virulence traits, efficient capture of nutrients and, enhancement of cell-to-cell communication.[4]

Both bacterial and yeast biofilm have been widely studied but less attention has been given for filamentous fungi (ff) biofilms.[5-7] Filamentous fungi are especially adapted for growth on surfaces, as evidenced by their absorptive nutrition mode, their secretion of extracellular enzymes to digest complex molecules, and apical hyphal growth.[8,9] Despite the fact that the term 'biofilm' is rarely applied to ff, there are several descriptions indicative of biofilm formation in different medical, environmental and industrial settings.[10-14]

The occurrence of fungal water biofilms can be a source of taste, odour and visual appearance problems resulting in poor drink water quality.[15] Moreover, ff in potable water distribution systems may have direct effects on human health (allergenic or toxigenic species), contribute to the occurrence of nosocomial infections in immune-compromised individuals and contaminate foodstuffs during processing or preparation.[8,16] Microbially-induced corrosion, loss of indicator organism utility and the persistence contamination in water can be problems related with the development of biofilms.[17,18] Water systems worldwide have been shown to be colonized with pathogenic filamentous fungi.[15] Although in recent years studies of fungi in drinking water have received attention, detailed researches of ff biofilms on water are rare.[19-24]

Water for human consumption is often disinfected before it goes to the distribution system to ensure that potential microbial pathogens are inactivated. Chlorine, chloramines or chlorine dioxide are most often used because they are very effective disinfectants.[17] In contrast, biofilms can protect microbes from disinfectants and allow microbes injured by environmental stress and disinfectants to recover and grow. Moreover, biofilms react with chemical disinfectants reducing their availability for inactivating pathogens in the water.[25] As a matter of consequence, biofilms can be considered one of the reasons for persistent microbial contamination of the water. Conventional water treatment (coagulation/flocculation, filtration, and chlorination) can be effective in removing microfungal contaminants from water but a possible re-contamination can occur if supplementary chlorination of all water service reservoirs is not routinely carried out.[26]

The objective of this work was analyse the susceptibility of *Penicillium brevicompactum* biofilms and its single spores against free chlorine, the most common disinfectant used routinely in water treatment.

2 MATERIAL AND METHODS

2.1 Spores

Penicillium brevicompactum (MUM 05.17) supplied by *Micoteca da Universidade do Minho* (MUM, Braga, Portugal) was chosen as a model as it is the most commonly filamentous fungi isolated from Portuguese tap water.[21] Spores were collected from a 7-day pure culture in malt extract agar (MEA: malt extract 20 g, peptone 5 g, agar 20 g, distilled water 1 l) at 25 ºC by adding 2 ml of distilled water to plate. The spore suspension was re-suspended and homogenised (Vortex) for 1 min before quantification using a Neubauer counter chamber. The suspensions were standardized by dilution with water to a final concentration of 10^5 spores/ml.

2.2 Biofilms

A spore suspension of 10^5 spores/ml was also used to perform biofilms. The biofilms were grown in 6-well plates at room temperature and 120 rpm. The spore suspension was added to each well which contained 5 ml of glucose solution (0.1 %). Then the PVC (polyvinyl chloride), PP (polypropylene) and PE (polyethylene) coupons (1 cm x 1 cm), previously autoclaved at 121 ºC during 15 min, were placed into the wells with the reverse face touching the well bottom and staying all under the water. After a time period 48h, 72h and 96h of incubation the biofilm on the coupons were used for free chlorine susceptibility test.

2.3 Concentrations of Free Chlorine

Sodium hypochlorite solutions were prepared with bleach and distilled and deionised water and adjusted to pH 7.0 ± 0.1 using HCl. For free chlorine determination, in the absence of iodide ion, free chlorine reacted instantly with DPD (*N,N*-diethyl-*p*-phenylenediamine) to produce a red colour, which was measured immediately with a colorimeter (Ion specific meters, Hanna Instruments, HI 93701, light emitting diode at 555 nm, range 0.00 to 2.5 mg/l, resolution 0.01 mg/l). The survival of *P. brevicompactum* spores and biofilms were determined after exposure to 0.015, 0.3, 0.6, 0.125, 0.25, 0.5 and 1% (v/v) of sodium hypochlorite, i.e., 0.02, 0.051, 0.25, 1.83, 1.98, 2.13 and 2.38 mg/l of free chlorine, respectively, during 15 minutes. A 10 % sodium hypochlorite solution was used as negative control. This dilution reflects the concentration currently used for surface sterilization.

2.4 Treatment with Free Chlorine

2.4.1 Spores. Pellets of 10^5 spores were re-suspended in 1 ml of free chlorine solution (Table 1), mixed by inversion to ensure full contact and then incubated at room temperature (25 ± 1 °C) for the required exposure time (15 min). The samples were mixed by inversion at least twice during incubation period, centrifuged and at the end the supernatants were discarded. Pellets were immediately washed with abundant distilled water and then centrifuged. This process was repeated three times. Finally, the resulting pellets were re-suspended in 1 ml distilled water. Positive controls were treated in similar way, with the exception that distilled water replaced the free chlorine solutions. In this study, free chlorine Minimum Inhibitory Concentration (MIC_{90}) is the concentration at which 90% of spores are inactivated after contact with the disinfectant solution (value expressed as a percentage of the colonies formed by positive control). For each concentration test and control, three replicates were made.

2.4.2 Biofilms. The PVC, PE and PP coupons were washed with distilled water to remove the non-adherent cells. Each coupon was transferred to another 6-well plate with different free chlorine solutions (0.02, 1.57 and 2.38 mg/l) and submerged. After 15 min the coupons were taken off and washed three times with distilled water to remove the disinfectant. All biofilms submitted to the different free chlorine concentrations and grown on different surfaces were done in triplicate. Both spores and biofilms were exposed to free chlorine solutions using the same conditions.

2.5. Spores Viability Test

2.5.1 Culture test. Spores viability was determined by their germination capability. The samples 1 ml volume of the spore suspension was plated using a pour plating method. To facilitate the determination of the number of viable spores, three 10-fold dilutions were prepared from each sample. A 1 ml volume of the spore suspension from the last dilution (i.e., maximum of 10^2 spores) was added to a sterile disposable Petri dish; 10 ml PDA (46 ± 2 °C) were then added and the mixture was gently swirled to evenly distribute the spores. The germination capability was confirmed after 72 h of incubation at 25 °C by the visible grown of colonies.

2.5.2 FUN-1 staining. P. brevicompactum spores without treatment were used to establish the method. The resulted images were used as standard for further comparison with treated spores (data not shown). From each free chlorine concentration tested, one spore suspension was chosen for FUN-1 staining (Molecular Probes, The Netherlands). FUN-1 stains the dead cells with a diffuse yellow-green fluorescence and the metabolically active cells with red Cylindrical Intra-Vacuolar Structures (CIVS). For FUN-1 staining, 15 μL of the spore suspension plus 15 μL of FUN-1 solution were added on a glass slide, homogenised, following incubation in the dark at 30 °C during 30 min and observed under an Olympus BX51 epifluorescence microscope using UV light equipped with 40x/0.30 and 10x/0.65 objectives and a filter set (EX 450-490 nm, EM 520). The images were acquired with a colour camera Zeiss AxioCam HRc using the software CellB®. Storage and handling of reagents were performed as recommended by the supplier.

2.6 Biofilm Viability Test

2.6.1 Culture test. The biofilms recovery was determined by plating the coupons on potato dextrose agar (PDA) plates. Before plating, the coupons were washed with distilled water and put on the culture medium with the biofilm touching the culture medium surface. The development of colonies was observed until 72 h of incubation at 25 ºC.

2.6.2 FUN-1 staining. From each free chlorine concentration tested, one coupon was chosen for FUN-1 staining. Negative and positive controls were also submitted for *in situ* viability test with FUN-1 staining as described in section 2.5.2.

3 RESULTS

Table 1 shows the sodium hypochlorite concentrations and the corresponding nominal concentrations of free chlorine. All disinfectant data refer to soluble free chlorine concentrations. The results after treatment with different free chlorine concentrations against spores and biofilms are shown in Table 2 and 3, respectively.

Table 1 *Sodium hypochlorite and soluble free chlorine concentration used in the present study.*

Sodium hypochlorite concentration (%, v/v)	Free chlorine nominal concentration (mg/l)
1.00	2.38
0.50	2.13
0.25	1.98
0.12	1.83
0.60	0.25
0.30	0.05
0.02	0.02

Table 2 *Number of spores of* P. brevicompactum *recovered after contact with different concentrations of free chlorine*; **each value is the mean of three independent assays (each assay with three replicates). Values expressed as a percentage of the colonies formed by positive control.*

Free chlorine (mg/l)	Spores recovered* (%)
2.38	0
2.13	0
1.98	0
1.83	8.2
0.25	23.7
0.05	98.4
0.02	99.5

3.1 Germination Capability

3.1.1 Spores. Low concentrations of free chlorine (0.02 and 0.05 mg/l) had no effect on the inactivation of *P. brevicompactum* spores. In contrast, high concentrations (1.98, 2.13 and 2.38 mg/l) give non-viable spores under the present experimental conditions. The number of fungal colonies observed on dilution plates derived from each treatment were enumerated 3 days post inoculation. No new colonies were observed 5 days post inoculation, thereby confirming that slow developing spores were not overlooked in the earlier counts. The MIC_{90} of free chlorine is 1.26 mg/l. Negative control inactivated all spores. Positive control gave an average of 2.5 x 10^2 spores recovered per ml.

Table 3 *Survival of* P. brevicompactum *biofilms determined by germination capability (+) after contact with different concentrations of free chlorine.*

Free chlorine (mg/l)	Biofilms								
	PVC			PE			PP		
	Biofilm age								
	48h	72h	96h	48h	72h	96h	48h	72h	96h
0.07	+	+	+	+	+	+	+	+	+
1.57	+	-	+	-	+	+	+	+	-
2.38	+	-	+	-	-	+	-	-	-
Positive control	+	+	+	+	+	+	+	+	+
Negative control	-	-	-	-	-	-	-	-	-

3.1.2 Biofilms. The lowest free chlorine concentration did not show any significant activity after 15 min exposures. For 0.07 mg/l of free chlorine, biofilms with 48, 72 and 96 h aged showed similar results with development of visible colonies after 48 h of incubation. For the intermediary free chlorine solution (i.e., 1.57 mg/l), most of biofilms were resistant and showed visible colonies until 72 h of inoculation. For the highest free chlorine concentration (i.e., 2.38 mg/l), age and material coupons interfered in the biofilm resistance; 96 h aged biofilms on PVC and PE were resistant, whereas none at 48 h and 72 h aged nor biofilms on PP presented colonies grown until 72 h of incubation.

The effects of free chlorine solutions in the deactivation of biofilms seem to be more related with the delay of the development of visible grown than with the inactivation itself. For example, for the 48 h aged biofilms on PVC, all free chlorine solutions tested did not inactivate the cells but the recovery time was different for each solution.

3.2 FUN-1 staining

3.2.1 Spores. After 30 min of incubation with FUN-1 viable spores from a 7-day pure culture were detected by conversion of FUN-1 dye into bright orange-red CIVS. The positive results allowed the comparison with treated spores. No evidence of autofluorescence was recovered in unstained spores. Effectiveness of free chlorine solutions against biofilms could be analysed by the FUN-1 staining under the conditions presented in this work.

For the highest free chlorine concentration, the spores were inactivated and did not form CIVS. The visualization of only diffuse green to green-yellow cytoplasmic staining indicates a membrane-compromised dead cell. After exposure to the lowest free chlorine concentration viable spores shown, in addition to diffuse green to green-yellow cytoplasmic staining, CIVS which had distinct orange-red fluorescence. For the intermediary free chlorine concentration both viable and non viable spores were detected (data not shown). The results could be correlated qualitatively with conventional plating and was less time consuming.

3.2.2 Biofilms. Viable biofilms were detected after 30 min of incubation with FUN-1 by conversion of FUN-1 dye into bright orange-red CIVS (Figure 1-B). The results allowed the comparison with treated biofilms. We investigated whether FUN-1 can be used to detect viability after exposure to free chlorine solution. In general, the intensity of fluorescent signals was lower in the treated biofilms when compared with non treated biofilms. Nevertheless, the results obtained with treated biofilms after FUN-1 staining were conclusive under the conditions presented in this work.

Effectiveness of free chlorine solutions against biofilms could be analysed by the FUN-1 staining under the conditions presented in this work. Although the results did not allow quantitative analyses, the qualitative results were conclusive. *P. brevicompactum* biofilms took up and converted the FUN-1 dye into bright orange-red CIVS. Free chlorine susceptibility test compared with conventional plating is more rapidly determined by viability analysis with FUN-1. The results suggest that analysis of biofilm viability with a fluorescent probe provides rapid and reproducible detection of cell inactivation. FUN-1 staining needs 30 min of incubation, whereas cultivation needs at least 48 h of incubation. Instead of the better development of biofilms on PVC coupons, its autofluorescence affected fluorescent signals and the image analysis making this material not the most recommended. By the other hand, PP and PE coupons did not show this problem and they are recommended for laboratorial biofilm development and fluorescence analysis.

4 DISCUSSION

Many studies about filamentous fungi from water have been published in the last years but there is still a lack of information about filamentous fungi biofilms from water distribution system. Moreover, there are also few studies examining the effects of free chlorine on filamentous fungi biofilms from water, thence the results presented in this work are discussed mainly with studies published about bacterial biofilms. In this study, the relationship between the attachment of filamentous fungi to surfaces and disinfection with hypochlorite solution was analysed with *P. brevicompactum* biofilms. Biofilms on different surfaces and different aged were tested with crescent free chlorine solutions. Free spores were also tested. Pour plating method and FUN-1 staining were used for viability tests.

The results indicate that attachment of spores to surfaces and the development of a biofilm provide features for fungi to survive disinfection. Even in a 10^5 spores/ml suspension, spores were vulnerable to the lowest free chlorine concentration, whereas biofilms were resistant to the highest. Attachment to a surface alters the way a disinfectant interacts with a microorganism and its efficacy may also be unsatisfactory against pathogens within flocks or particles, which protect them from disinfectant action.[27,28] The free chlorine acts within the cell membrane inactivating microorganisms indicating that resistance against chlorine is linked with structural features provided by biofilms. The negatively charged of EPS are also efficient in protecting cells from positively charged biocides by restricting their permeation through binding. Additionally, a small portion of cells (persisters) could survive the common causes of cell death by the induction of quiescence in certain biofilm pockets. Such quiescent cells are noted for their resistance to biocides.[29] In fact, microorganisms growing attached to surfaces often display a distinct phenotype that provides resistance to biocides.[30-32]

There are previous studies showing that the characteristics of the pipe material can influence the formation of bacterial biofilms and the survival of pathogens in drinking water.[29,33-35] Biofilms 96 h aged on PVC surface were more resistant when compared with same aged biofilm on PP and PE surfaces. According to LeChevallier *et al.* (1990) increase disinfection efficiency is not based solely on disinfection concentration, i.e., two times more disinfectant concentration does not result in twice inactivation.[36] Biofilms on PVC showed a different structural composition with an organized mycelial development and a mature structure. Biofilms on PP and PE surfaces showed cells sparsely distributed and were less resistant against chlorine inactivation suggesting that more than cellular density the biofilm architecture can difficult the transport of the disinfectant to the biofilm interface increasing the chlorine demand for cell inactivation. Lehtola *et al.* (2004) compared biofilm formation on cooper and PE surface and found that biofilms had a different rate of development and a different microbial community structure but that after one year microbial numbers in biofilms in water were similar in both materials.[37]

Often household plumbing is constructed of plastic or copper, in some certain cases of stainless steel. Into household plumbing there are increases in the temperature and the content of chlorine decreases and consequently, microbial numbers increase in the water distributed throughout the buildings.[38,39] In this study, for filamentous fungi biofilm, PVC was the best surface for its development. In contrast, Yu *et al.*, (2010) indicate some plastic materials, such as PVC, for drinking water distribution pipes, due to its low biofilm formation potential and little microbial diversity in biofilm.[40]

In a research with filamentous fungi, Ramírez-toro *et al.* (2002) showed that organisms were able to colonise glass slides even in the presence of chlorine concentrations higher than those normally found in distribution systems.[41] These authors also found that older biofilms are more resistance to chlorine and that attachment allowed survivals 2 to 10 times higher than planktonic cells. The highest concentration of chlorine used in this work was greater than the concentration advised by the World Health Organization (between 0.2 and 0.5 mg/l).[28] The antimicrobial activity of chlorine depends on the amount of hypochlorous acid which, in turn, depends on the pH, the amount of organic matter and on the temperature of water. However, excessive treatment, i.e. hyperchlorination, has several known and potential negative effects on product sensory quality, in environment and human health. Moreover, disinfection with free chlorine can also be affected by pipe surface and biofilm age.[27]

The efficiency of disinfection is important to reduce microorganisms in water and to avoid contamination. According to Council Directive 98/83/EC, a 0.5 mg/l concentration of free chlorine, after an exposure of 30 min, guarantees a satisfactory disinfection. The already published researches about biofilms disinfection showed that this free chlorine concentration is not effective against bacterial biofilms. In an overview, the water distribution network is under many different changes such as pH, nutrients, pipes material, temperature that influence in chlorine effectiveness what make disinfection a complex step of water treatment.

The study of biofilms has increased in the last years and the development of new methodologies has a great importance. Biofilms from water distribution system are known as a resource of microorganisms and recontamination with a consequently reduction in water quality. FUN-1 staining associated with fluorescence microscopy is a non destructive analytical technique which in association with others fluorescent dyes provides metabolic and morphological analyses.

FUN-1 has been widely applied for antifungal susceptibility tests. Balajee and Marr (2002) reported a flow cytometric assay relied on conidial metabolism of the viability dye FUN-1 with spores of *Aspergillus* sp.[42] Susceptibility of *Candida* sp. clinical isolates was also investigated with FUN-1 staining and the authors suggested it as an alternative and rapid method emphasize that the use of fluorescent viability assays can indicate the presence of a viable but not cultivable spores state.[43,44] De Vos *et. al.* (2006) used fluorescent dyes to detect fungi in hospital waters, dialysis fluids and endoscopic rinse. The results of these authors shown that in general, fluorescent labelling techniques detected more fungi in water than plate methods.[45]

In this study, FUN-1 staining didn't allow a quantitative analyse since a biofilm, as its own definition already explains, is not compose of free cells but the qualitative analyses were conclusive and the results corresponded with conventional plating results. The results already published are most about detection of yeast and spores viability and the application of FUN-1 dye for filamentous fungi biofilm are few or were not published yet.

In conclusion, we presented a simple and reproducible methodology for the study of the effectiveness of free chlorine against filamentous fungi biofilms from water. For this, we applied conventional plating and FUN-1 staining and we have shown that FUN-1 is efficient and offered rapid and reliable results for laboratorial biofilms and more studies are necessary to apply the methods in real biofilms. Furthermore, this is the first report about FUN-1 staining for susceptibility and viability analyse of a filamentous fungi biofilms from water. Finally, *P. brevicompactum* biofilms were capable to survive after exposure to a high free chlorine concentration whereas free spores were susceptible.

References

1 G. O'Toole, HB Kaplan, R Kolter, *Annu. Rev. Microbiol.*, 2000, **54**, 49–79.
2 J. Chandra, DM Kuhn, PK Mukherjee, *J Bacteriol*, 2001, **183**, 5385-5394.
3 J.J. Harrison, R.J. Turner, and H. Ceri, *American Scientist*, 2005, **93**, 508-515.
4 M.W. Harding, L.R.L. Marques, R.J. Howard and M.E. Olson, *Trends in Microbiol.*, 2009, **17** (11), 475-480.
5 G.A. O'Toole, L.A. Pratt, P.I. Watnick, D.K. Newman, V.B. Weaver and R. Kolter, *Methods Enzymol.*, 1999, **310**, 91–109.
6 P. Watnick and R. Kolter, *J. Bacteriol.*, 2000, **182**, 2675–2679.
7 A. Huq, C.A. Whitehouse, C.J. Grim, M. Alam and R.R. Colwell, *Curr. Opin. Biotechnol.*, 2008, **19**, 244–247.
8 K.T. Elvers, K. Leeming and H.M. Lappin-Scott, *J Ind Microbiol Biotechnol*, 2001, **26**, 178-183.
9 G.K. Villena, T. Fujikawa, S. Tsuyumu and M. Gutiérrez-Correa, *Bioresour. Technol.*, 2009, doi:10.1016/j.biortech.2009.10.036.
10 M.S. Doggett, *Appl. Environ. Microbiol.*, 2000, **66**, 1249–1251.
11 A. Beauvais, C. Schmidt, S. Guadagnini, P. Roux, E. Perret, C. Henry, S. Paris, . Mallet, M-C Prévost, J.P. Latgé, *Cell. Microbiol.*, 2007, **9**, 1588–1600.
12 R. Paterson, A. Gonçalves and N. Lima In: *Proceedings of the 8th International Mycological Congress*. SAPMEA, Eastwood, Australia, 2006, 129.
13 E. Mowat, C. Williams, B. Jones, S. Mcchlery, G. Ramage. *Med. Mycol.*, 2008a, **47** (1), S1–S7.
14 E. Mowat, S. Lang, C. Williams, E. McCulloch, B. Jones and G. Ramage, *J. Antimicrob. Chemother.*, 2008b, **62**, 1281–1284
15 G. Hageskal, N Lima, I. Skaar, *Mycol. Res.*, 2008, doi:10.1016/j.mycres..10.002.
16 E.J. Anaissie, S.L. Stratton, M.C. Dignani, C-k Lee, R.C. Summerbell, J.H. Rex, T.P. Monson and T.J. Walsh, *Blood*, 2003, **101**, 2542–2546.
17 U.S. EPA. U.S. Environmental Protection Agency. Drinking water criteria document on heterotrophic bacteria. Washington, DC. 1984.

18 E.E. Geldreich. In: *Microbial quality of water supply in distributions systems*. Lewis Publishers, Boca Raton, FL, 1996.

19 E. Göttlich, W. van der Lubbe, B. Lange, S. Fiedler, I. Melchert, M. Reifenrath, H.-C. Flemming, S. de Hoog, *Int J Hyg Environ Health*, 2002, **205**(4), 269-79

20 G. Hageskal, AK Knutsen, P Gaustad, GS de Hoog, I Skaar, *Appl Environ Microbiol*, 2006, **72**, 7586–7593.

21 AB Gonçalves, RRM Paterson, N Lima, *Int J Hyg Environ Health*, 2006, **209**, 257-264.

22 A Ribeiro, AP Machado, Z Kozakiewicz, M Ryan, B Luke, AG Buddie, A Venâncio, N Lima, J Kelley, *Rev Iberoam Micol*, 2006, **23**(3): 139-144.

23 M.U. Yamaguchi, R.C.P Rampazzo, S.F. Yamada-Ogatta, C.V. Nakamura, T. Ueda - Nakamura, B.P.D. Filho, *Braz Arch Biol Technol.*, 2007, **50** (1), 1-9.

24 V.J. Pereira, M.C. Basílio, D. Fernandes, M. Domingues[a], J.M. Paiva, M.J. Benoliel, M.T. Crespo, *Water Res.*, 2009, **43**, 3813 – 3819.

25 P.S. Berger, R.M. Clark and D.J. Reasoner, In: *Encyclopedia of Microbiology*. 2nd Edition., 2000, **4**, 898-913.

26 N.B. Sammon, K.M. Harrower, L.D. Fabbro and R.H. Reed, *Int. J. Environ. Res. Public Health*, 2010, **7**, 1597-1611

27 M.W LeChevallier, C.D. Cawthon, R.G. Lee, *Appl Environ Microbiol*, 1988, **54** (3), 649-654.

28 World Health Organization. *Guidelines for drinking water quality*, vol 1. World Health Organization, Geneva, 2008.

29 T. Schwartz, S. Hoffmann and U. Obst, *Appl. Microbiol.*, 2003, **95**, 591–601.

30 R. Srinivasan, P.S. Stewart, T. Griebe, C.I. Chen, X. Xu, *Biotechnol Bioeng*, 1995, **46**(6), 553-560.

31 W.L. Cochran, G.A. McFeters, P.S. Stewart, *J Appl Microbiol*, 2000, **88**(3), 22-30.

32 J. Morato', F. Codony, J. Mir, J. Mas, F. Ribas, Microbial Response to Disinfectants. *In*: Mara, D., Horan, N. (Eds.), *The Handbook of Water and Wastewater Microbiology*. Academic Press, London, 2003, p.657-693.

33 P. Niquette, P. Servais, R. Savoir, *Water Res.*, 2000, **34**(6), 1952–1956.

34 C.D. Norton, M.W. LeChevallier, J.O. Falkinham III, *Water Res.*, 2004, **38**(6), 1457–1466.

35 M.J. Lehtola, I.T. Miettinen, T. Myllykangas, A. Hirvonen, T. Vartiainen, P.J. Martikainen, *Water Res.*, 2005, **39**, 1962–1971.

36 M.W. LeChevallier, C.D. Lowry, R.G. Lee, *J Am Water Works Assoc*, 1990, 82(7), 87-99.

37 M.J. Lehtola, I.T. Miettinen, M.M. Keinanen, T. Kekki, O. Laine, A. Hirvonen, T. Vartiainen, P.J. Martikainen, *Water Res.*, 2004, **38**, 3769–3779.

38 O.M. Zacheus and P.J. Martikainen, *Sci. Total Environ*, 1997, **204**, 1–10.

39 O.M. Zacheus and P.J. Martikainen, *Can. J. Microbiol.*, 1995, **41**, 1088–1094.

40 J. Yu, D. Kim, T. Lee, *Water Sci Technol.*, 2010, **61**(1), 163-71.

41 G.I. Ramírez-toro and H.A. Minnigh, In: *XXVIII Congreso Interamericano de Ingeniería Sanitaria y Ambienta*, Cancún, México, 2002.

42 S.A. Balajee1 and K.A Marr, *J Clin Microbiol*, 2002, **40**(8), 2741–2745.

43 C. Pina-Vaz, F. Sansonetty, A.G. Rodrigues, S. Costa-de-Oliveira, J. Martinez-de-Oliveira and A.F. Fonseca, *J. Med. Microbiol.*, 2001, **50**, 375-382.

44 L.M.E. Vanhee, H.J. Nelis and T. Coenye, *J Microbiol Methods*, 2008, **72**, 12–19.

45 M.M. De Vos and H.J. Nelis, *J Microbiol Methods*, 2006, **67**, 557–565.

NEAR REAL TIME MONITORING OF *E. COLI* IN WATER

F. Zibuschka [1], T. Lendenfeld [2], G. Lindner [1]

[1] Institute for Sanitary Engineering and Water Pollution Control, University of Natural Resources and Life Sciences, Vienna (BOKU), Muthgasse 18, A-1190 Vienna, Austria
[2] mbOnline GmbH, Steiner Landstraße 27a, A-3500 Krems, Austria

1 INTRODUCTION

There is a need for microbiological testing methods that provide results faster than the standard methods. Unlike other areas of microbiological diagnostics new robust rapid detection methods for the water engineering profession are not widely available. This is mainly because the conventional microbiological analysis techniques are established worldwide via the respective national and international standards. To be faster than standard cultivation methods means omitting the step of cultivation, which consequently leads to detection of enzymatic activity in the sample, which should be sufficient to achieve a reliable and sensitive result. These considerations required the design of a reactor-chamber combined with a filtration unit to enrich the bacteria prior to the enzymatic assay stage. After a proof-of-concept, a prototype was developed and then made ready for field use. One of the first applications of the device was the Near Real Time Monitoring of *E. coli* in the effluent of three membrane filtration units of a pilot plant for waste water treatment.

2 PREFACE

The outcome of this is on the one hand, the desired result of comparability and fitness of purpose of the data, on the other hand, however, the disadvantage of a multi-day waiting period on the test results with conventional methods. It follows a delayed identification of possible problems and from this a delayed taking of the necessary remedial actions. A large number of methods of microbiological investigations permit greater differentiation in studies for determining water quality, which enhances the overall safety of assessment. Novelties in the field of microbiological analytics promote the development of alternative methods based on different measuring techniques. In the present case, a "Near-Real-Time" process is presented and first results are reported.

3 MATERIALS AND METHODS

To be faster than standard cultivation methods means the rejection of the step of cultivation, which consequently leads to detection of enzymatic activity in the sample, which should be sufficient to gain a reliable result. These considerations ended up in the design of a reactor chamber combined with a filtration unit to enrich bacteria prior to the enzymatic assay. In combination with a highly fluorescence detection unit the metabolism could be detected. After a proof-of-concept, a prototype was developed and ready for field use. This prototype was further improved and gained serial production status in 2009 (COLIGUARD® EC. hs, Figure 1).

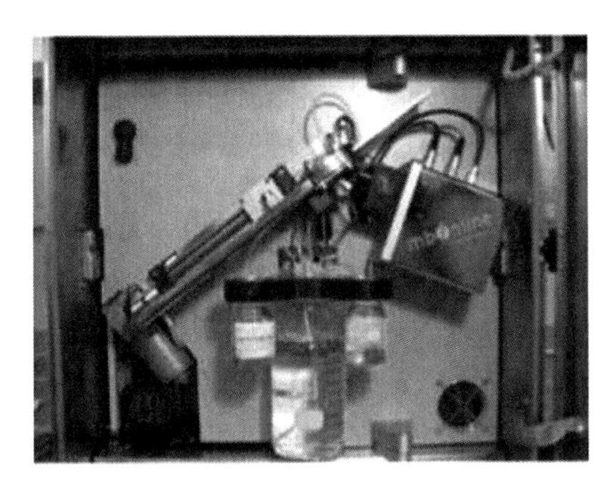

Figure 1 *COLIGUARD® EC.hs, Reactor with filter plate, fluorescence optical measuring unit, below reagents and cleaning solution*

4 RESULTS

One of the first applications of the device was the Near Real Time Monitoring of *E. coli* in the effluent of three membrane filtration units of a pilot plant for waste water treatment (See Figures 2 and 3). The pilot plant was operated using a combination of a conventional sequencing batch principle (SBR) with submerged flat membrane modules (MBR), namely one ultrafiltration membrane unit MICRODYN NADIR GmbH, Type Bio-Cel®-BC-10-10-PVC, pore size 0,04 µm (UFU) and two microfiltration membrane units, ItN Nanovation AG, Type A-HP, pore size 0.2 µm (MFU 1); Kubota, Type M-Box, pore size 0.4 µm (MFU 2).

The aim of the project was to treat wastewater to such an extent that a safe infiltration into the subsoil is possible. To check whether a secure infiltration of treated wastewater is possible, the effluents of the three membrane units were investigated by COLIGUARD® EC.hs. Since only one COLIGUARD® EC.hs was available the effluents could not be studied simultaneously. Therefore the investigations had to be carried out sequentially. The results of the Near Real Time Monitoring of the effluents of the tree membrane units are summarized in Figure 3. The arrows mark the examination time for each membrane unit.

The investigations showed that the ultrafiltration membrane was damaged and it therefore led to breakthroughs of *E. coli*. In comparison the two microfiltration modules resulted in significantly better effluent quality. The effluent of MFU 1 had the lowest *E. coli* contamination. In this context it should be noted that a comprehensive assessment of the effectiveness of a treatment plant can be done only with a Near Real Time Monitoring since only in this way the required data can be preserved. Using standard microbiological tests it is hardly possible to examine several samples a day to get the assessment data and still do this over a longer period of time. The advantage of the Near Real Time Monitoring compared with microbiological standard procedure is, that short-term changes of water quality (gradient graph) are recognizable within few hours. In many cases, short-term changes in water quality can be seen only with this kind of technique. Thus, the knowledge about the actual water quality increases and more confidence in operating the plant will be achieved.

5 CONCLUSION

The COLIGUARD® EC.hs is well suitable for monitoring *E. coli* contamination of water. The sensitivity makes the system an interesting tool for monitoring microbiological contamination. The system meets the need for "rapid detection for rapid decisions".

Figure 2 *Pilot plant, black feed tank, grey permeate tanks of UFU (1), MFU 1 (2) and MFU 2 (3)*

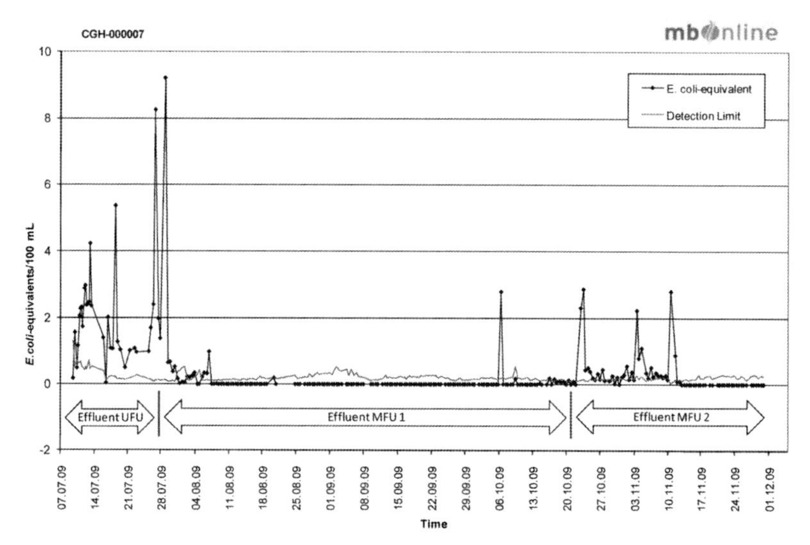

Figure 3 *E. coli, effluent of different membrane filtration units*

Acknowledgement
This work was supported by the Government of Austria, Federal Ministry of Agriculture, Forestry, Environment & Water Management.

Subject Index